How Wars End

This book addresses one of the most important issues in international relations – how wars are ended.

The volume draws on the direct experience of both soldiers and academics, who in each case have also been advisers on fighting and ending wars. Unlike more theoretical works, the book draws on first-hand experiences in the case studies, which include the wars in Iraq, Afghanistan, Sierra Leone and Indonesia, among others. The volume is constructed around a series of themes. The first theme is why wars start and how they can be understood, based on the assumption that knowing how, and why, wars start is fundamental to understanding how they might end. The second is what sustains wars and what makes them difficult to end. Again, once wars start, understanding what keeps them going is critical to how to end them. The third focuses on the role of external intervention in ending wars, including as a belligerent partner in war, as a peacemaking or peacekeeping force, and as a mediator between warring parties. The fourth addresses the issue of 'ripeness' and the right conditions for ending wars. The fifth addresses the modalities for ending wars and creating peace, with the sixth theme being focused on transitions to peace and what is required to help make those transitions successful.

The book will be of interest to students of military, strategic and security studies, peace studies and International Relations.

Richard Iron CMG OBE is the president of the Victorian Branch of the Australian Institute of International Affairs. He has served in the British Army, was Defence Fellow at the University of Oxford and worked for the UK's Chief of Defence Staff on strategic planning.

Damien Kingsbury is an emeritus professor with the School of Humanities and Social Sciences at Deakin University, Melbourne. He was the principal adviser to the Free Aceh Movement in the 2005 Helsinki Peace Talks and has advised a number of other armed non-state groups in conflict resolution. He is the author or editor of more than two dozen books, most recently including *Separatism and the State*.

Routledge Advances in Defence Studies

Series Editors:

Timothy Clack

University of Oxford, UK

Oliver Lewis

Rebellion Defence and University of Southern California, USA

Routledge Advances in Defence Studies is a multi-disciplinary series examining innovations, disruptions, counter-culture histories, and unconventional approaches to understanding contemporary forms, challenges, logics, frameworks, and technologies of national defence. This is the first series explicitly dedicated to examining the impact of radical change on national security and the construction of theoretical and imagined disruptions to existing structures, practices, and behaviours in the defence community of practice. The purpose of this series is to establish a first-class intellectual home for conceptually challenging and empirically authoritative studies that offer insight, clarity, and sustained focus.

The World Information War
Western Resilience, Campaigning and Cognitive Effects
Edited by Timothy Clack and Robert Johnson

Making British Defence Policy
Continuity and Change in an Uncertain World
Robert Self

Strategic Autonomy and Economic Power
The Economy as a Strategic Theater
Vitor Bento

Cultural Heritage in Modern Conflict
Past, Propaganda, Parade
Edited by Timothy Clack and Mark Dunkley

How Wars End
Theory and Practice
Edited by Richard Iron and Damien Kingsbury

For more information about this series, please visit: https://www.routledge.com/Routledge-Advances-in-Defence-Studies/book-series/RAIDS

How Wars End

Theory and Practice

Edited by Richard Iron and Damien Kingsbury

Routledge
Taylor & Francis Group
LONDON AND NEW YORK

First published 2023
by Routledge
4 Park Square, Milton Park, Abingdon, Oxon OX14 4RN

and by Routledge
605 Third Avenue, New York, NY 10158

Routledge is an imprint of the Taylor & Francis Group, an informa business

British Library Cataloguing-in-Publication Data
A catalogue record for this book is available from the British Library

Library of Congress Cataloging-in-Publication Data
Names: Iron, Richard, editor. | Kingsbury, Damien, editor.
Title: How wars end: theory and practice / edited by Richard Iron and Damien Kingsbury.
Description: Abingdon, Oxon ; New York, NY : Routledge, 2023. | Series: Routledge advances in defence studies | Includes bibliographical references and index. |
Identifiers: LCCN 2022035494 (print) | LCCN 2022035495 (ebook) | ISBN 9781032329512 (hardback) | ISBN 9781032329529 (paperback) | ISBN 9781003317487 (ebook)
Subjects: LCSH: War. | War--Causes. | Peace.
Classification: LCC U21.2 .H6525 2023 (print) | LCC U21.2 (ebook) | DDC 355.02--dc23/eng/20220927
LC record available at https://lccn.loc.gov/2022035494
LC ebook record available at https://lccn.loc.gov/2022035495

ISBN: 978-1-032-32951-2 (hbk)
ISBN: 978-1-032-32952-9 (pbk)
ISBN: 978-1-003-31748-7 (ebk)

DOI: 10.4324/9781003317487

Typeset in Times New Roman
by MPS Limited, Dehradun

Contents

Preface vii
List of Contributors viii

Introduction 1
DAMIEN KINGSBURY AND RICHARD IRON

SECTION 1
Why War? 9

1 The Causes of War—and their Consequences 11
HEW STRACHAN

2 A Typology of Wars and How They End 28
ROB JOHNSON

3 How You Fight a War Matters to How It Ends: A Real-World Contemporary Case Study—Iraq 2016 45
ROGER NOBLE

SECTION 2
Case Studies 57

4 Libya's Mercenaries Crisis: 'I Am Here to Protect the King of Kings' 59
ALIA BRAHIMI

5 Indonesia's War Against East Timor: How It Ended 70
CLINTON FERNANDES

6 Ending the Sierra Leone War 89
RICHARD IRON

7 Peace Processes in Aceh and Sri Lanka: A Comparative Assessment 107
DAMIEN KINGSBURY

8 The Case of Afghanistan: How Wars End 123
DAVID KILCULLEN

9 Some Reflections on the Pursuit of 'Peace' in Afghanistan: 'Never Send to Know for Whom the Wars End' 139
WILLIAM MALEY

SECTION 3
Alternatives 155

10 Reflections on the Australian Experience: How Wars End 157
JOHN BLAXLAND

11 How Major-Power Wars End 171
HUGH WHITE

SECTION 4
Ways Forward 187

12 Negotiations to End all Wars 189
WILLIAM ZARTMAN

13 Negotiating Peace 202
DAMIEN KINGSBURY

14 Endless Wars, Perpetual Peacekeeping? 221
ADAM DAY AND CHARLES T HUNT

15 Conclusion 236
DAMIEN KINGSBURY AND RICHARD IRON

Index 249

Preface

The idea for this edited collection began as an after-dinner discussion between the incipient editors in 2018, and progressed through a presentation to the Australian Institute for International Affairs before an agreement to hold a (virtual) conference on the subject at the Institute and to prepare this volume.

The core of the subject of how wars end, and how to bring them to an earlier conclusion, has been central to the work of both editors, if in different capacities.

Richard Iron and Damien Kingsbury

Contributors

John Blaxland is a professor of International Security and Intelligence Studies at the Australian National University. He wasformer Director Joint Intelligence Operations and is a senior fellow of the Higher Education Academy, a Fellow of the Royal Society of New South Wales, and the first Australian recipient of a US Department of Defense Minerva Research Initiative grant.

Dr Alia Brahimi is a political advisor and a leading specialist in terrorism and political trends in the Middle East and North Africa. She was most recently a visiting fellow at the Oxford University Changing Character of War Programme. She was previously a research fellow at the London School of Economics and a Research Associate at the University of Oxford.

Dr Adam Day is the head of the Geneva Office of United Nations University Centre for Policy Research. He was previously senior political adviser to the United Nations Mission in the Democratic Republic of Congo and worked for the UN in Lebanon, Khartoum and Darfur, and the Department of Political Affairs and the Department of Peacekeeping Operations in New York.

Professor Clinton Fernandes, School of Humanities and Social Sciences, University of New South Wales, Australia, served in the Australian Army Intelligence Corps. He has published on the relationship between science, diplomacy and international law, intelligence operations in foreign policy, including Island off the Coast of Asia: Instruments of Statecraft in Australian Foreign Policy (Rowman and Littlefield, 2018).

Dr Charles Hunt is an associate professor of International Relations in the School of Global, Urban and Social Studies at RMIT University. He is a senior fellow at the United Nations University Centre for Policy Research in New York, and an honorary senior research fellow at the Asia Pacific Centre for the Responsibility to Protect at the University of Queensland.

Richard Iron CMG OBE served in the British Army in Northern Ireland, the Sultanate of Oman, the Falkland Islands, the Balkans and Iraq and was an expert military witness in the Sierra Leone War Crimes trials. He has been a Defence Fellow at the University of Oxford and has worked for the UK's Chief of Defence Staff.

Dr Rob Johnson, Director of the UK Office of Net Assessment, having been Director of the Oxford Changing Character of War Centre, senior research fellow of Pembroke College, and associate of the Department of Politics and International Relations, University of Oxford. He is also an adjunct professor at the Norwegian Defence University Staff College and the Strategic Studies at Rennes School of Business.

Damien Kingsbury is an emeritus professor with the School of Humanities and Social Sciences at Deakin University, Melbourne, where he formerly held a Personal Chair and was a professor of International Politics. He was the principal adviser to the Free Aceh Movement in the 2005 Helsinki Peace Talks and has advised a number of other armed non-state groups in conflict resolution.

Dr David Kilcullen FRGS is a professor at Arizona State University and at University of New South Wales, Canberra. He was chief strategist in the Office of the Coordinator for Counterterrorism at the U.S. State Department, special advisor for counter-insurgency to US Secretary of State Condoleezza Rice and key architect of the US 2007 Iraq 'surge' strategy.

Emeritus Professor William Maley AM was a professor of diplomacy at the Australian National University and was a visiting Professor at the Russian Diplomatic Academy, among other appointments. He is also a Barrister of the High Court of Australia, Vice-President of the Refugee Council of Australia, and a member of the Australian Committee of the Council for Security Cooperation in the Asia Pacific.

Major General (ret.) Roger John Noble, AO, DSC, CSC is the Australian Ambassador for Counter-Terrorism. He was deployed six times on operations to East Timor, Afghanistan and Iraq and seconded to the United States Army, serving as Deputy Coalition Land Force Commander, Iraq, as part of Operation Inherent Resolve, and as Deputy Commanding General – North in the United States Army Pacific.

Sir Hew Strachan is a professor of International Relations at the University of St Andrews, previously being Chichele Professor of the History of War at All Souls College, Oxford and director of the Changing Character of War program at the University of Oxford. He is the author of *The Direction of War: Contemporary Strategy in Historical Perspective* among other publications.

Hugh White AO is an emeritus professor of Strategic Studies at the Australian National University. He was an intelligence analyst with the Office of National Assessments, an adviser to Australian Prime Minister Bob Hawke and the deputy secretary for Strategy and Intelligence in the Australian Department of Defence, and the first director of the Australian Strategic Policy Institute.

William Zartman is a professor emeritus at the Paul H. Nitze School of Advanced International Studies of Johns Hopkins University and earlier directed the school's Conflict Management and African Studies programs. He holds the Jacob Blaustein Chair in International Organizations and Conflict Resolution and is a founder and current Board Chairman of the International Peace and Security Institute.

Introduction

Damien Kingsbury and Richard Iron

War is the scourge of our Earth. It is extraordinary that destruction and widespread killing of fellow humans has been and remains such a common feature of our behavior. Some ascribe this to something inherent in the human psyche; others to hyper-masculinity; some others to the idea that violence provides simple answers to complex questions. Yet war is the most unpredictable of human endeavors: the outcome very rarely matches that anticipated at war's outset. There is nothing simple about war.

It is even more extraordinary, given the human tragedy caused by war, that ending wars has proved so difficult; some wars extend for years, their original causes are long forgotten. Yet all wars do end eventually: we must learn from this bitter experience and hopefully understand better how to bring future wars to an end.

Why War?

War has, historically, been such a common feature of social behavior that one might be forgiven for thinking of it as the default position in human affairs. Certainly, some people and their political leaders are more inclined than others to resolve claims they might have by force of arms. For some, wars are in Clausewitzian terms a 'rational instrument of national policy' (Rapoport 1982: 13) and should, in an ideal world, be conducted rationally. Yet wars are usually not engaged in for purely rational reasons and, once engaged, rarely process along 'rational' lines. For the people who actually fight wars, or are the victims of wars, arguing war's rationality immediately appears a very thin proposition.

In this respect, there might be a question of how often war has reflected a conscious shaping of debate in ways which leave, or appear to leave, few alternatives. There are, of course, always alternatives to wars, even if they require more intellectual effort than simply engaging in armed violence. But wars have, and can still, appear as 'rational' to decision-makers who are not obliged to fight them and who are rarely their victims.

It might be simple, perhaps too simple, to suggest that humans have a predilection for war, given its recurrence. More likely, humans have a

DOI: 10.4324/9781003317487-1

predilection for simple answers to complex questions, with engaging in war being perhaps the greatest method of simplifying such complexity, even though it immediately raises a new set of its own, rather more challenging and often existential, complexities. Similarly, social anger or indignation can result in an impetus toward war; all that requires is a political leader ready to seize (or manufacture) the opportunity and to lead the metaphorical charge.

The reasons for people to go to war have, historically, not greatly changed, although the personal interests and pride of absolute rulers has largely diminished. Even so, political insults can escalate pre-existing tensions and be understood as betraying a deeper material antagonism. A people, if understanding themselves to have been insulted as a whole, can quickly rally behind a leader otherwise intent on belligerence. China and Japan, by way of illustration, have a history of derogatory language toward each other, recently reflected in China's claims to the Senkaku/Diaoyu Islands northeast of Taiwan. There is no doubt that undiplomatic language is unhelpful in avoiding an escalation of tensions, but it is fair to say that there is usually a deeper and more substantive issue also at play in such disputes.

The idea of empire has also been a driver of wars, with territorial expansion often couched in terms of 'civilizing mission'. In the post-imperial era, that language has been modified to call upon the removal of a malignant influence. To illustrate, President Wilson argued for the US involvement in World War I by arguing that 'the world must be made safe for democracy', the Cold War was littered with arguments in favor of liberation as well as the defeat of communism, the US-led invasion of Iraq was couched in terms of removal of a generic threat of 'weapons of mass destruction', and Russia's 2022 invasion of Ukraine was couched in terms of 'de-Nazification'.

Despite the sometimes lofty (or shallow rhetorical) claims made for them, many wars continue to be embarked upon with the intended goal of furthering national interests or enhancing or securing economic security. Historically, when people were a critical economic asset, wars were conducted to increase subservient populations. With industrialization, that emphasis changed to access to land and resources. Wars have even been initiated under the rubric of pre-emptive defense, such as Japan's 1941 attack on the US naval base at Pearl Harbor, Germany's invasion of Norway, or the Arab-Israeli Six Day War (Fearon 1995: 379–414). It was the logic of preparing for each other's assumed aggression that led to a European arms race and the readying of armies throughout Europe that primed Europe for World War I.

More credible is a war for self-defense, in the face of actual rather than prefigured attack. Ukraine is fully committed to its war with Russia, following Russia's invasion. In a related sense, that war is also over issues of sovereignty and territorial integrity in the face of what amounts to Russia's attempted territorial expansion.

The claim to territorial integrity is also used to justify a more local war when addressing threats from separatist movements. Yet separatists also legitimize their actions with claims to national identity in relation to a specific territory, and arguments in favor of self-determination ('autonomous development'), which were given initial substance by Wilson 1918 '14 Points' (Points X, XIV) and led to the realization of newly independent states in Europe (see also UN Charter 1;2).

Since 9/11, we have also been used to characterize war in terms of religious ideology, which appeals to competing world views and can sometimes obscure a more basic ethnic division or separatist identity. However, by invoking calls to religious radicalism, many of the legal and moral constraints developed to limit the conduct of war are effectively neutralized; as a result, wars of religion are frequently characterized by their extreme violence and high levels of non-combatant casualties.

The Tragedy of War

Joseph Stalin supposedly said that the death of one person is a tragedy, but the deaths of a million are a statistic. If he said this, he was wrong. The premature death of a person is indeed a tragedy, taking a life of potential and, as anyone who has seen war up close knows, turning it into lifeless flesh, of words left unsaid and aspirations irretrievably ended. There is nothing more absolute than death, nor less able to be appealed. The deaths of a million, therefore, is not a statistic but unfulfilled words and deeds multiplied by enormity: a million families shattered, a million times the grief, perhaps more so as a whole society in grief is reflected in intergenerational trauma.

Historically, war has frequently involved whole societies and sometimes resulted in widespread slaughter of combatants and non-combatants alike. Defeat of a civilization often meant its destruction, as well as its subjugation, such as after the Mongol Siege of Baghdad in 1258 and the Fall of Constantinople in 1453. During the European Age of Enlightenment, the conduct of war was more regulated, armies professionalized and civilians largely protected from the excesses of war. Yet, this was to change again with the revolutionary fervor unlocked by the French Revolution and, later, by the industrialization of armies and their weaponry, extending their reach and lethality and blurring the distinctions between combatants and non-combatants. Despite attempts to regulate the conduct of war through legal protocols, such as the Geneva Conventions, and protect non-combatants, civilians regularly and increasingly make up the larger portion of wars' casualties. Civilian casualties have risen from being a minority prior to the twentieth century to the point where, by the end of the twentieth century, non-combatant deaths regularly comprised up to 90% of total war deaths (Kaldor 1999: 107).

But the tragedy of war is more than personal tragedy writ large. War, no matter how localized, nearly always has greater and wider impact. Within a globalized economy, war may well interrupt trade including, potentially, imports of energy and food that affect vulnerable people far away from the conflict.

Given the likelihood of non-combatants becoming casualties of war, widespread fighting almost inevitably results in large flows of refugees and displaced people, creating strains on international organizations and potential host countries. And there is always potential for geographic escalation of war: regardless of how local the causes of war may be, other actors frequently see their own interests either threatened or furthered by war and are tempted to intervene. Through intervention, third parties become involved in the war and may create international blocks in opposition to each other, creating international fracture lines far greater and potentially more serious than the original war.

If the purpose of human society is the general advance of most, then war is its opposite. Where progress and development build, war destroys; where peaceful society values life and looks to its enhancement, war reduces life to a grim ledger and, at best, its reduction to the most utilitarian object of survival. War is anti-life, anti-progress and anti-development. It is, in its most fundamental sense, not so much 'policy by other means' as a failure of policy, a failure of political imagination and political and diplomatic skill; in short: war is the result of a cascade of mistakes. Ending wars, without absolute victory or loss, is therefore about correcting those mistakes or finding ways to ameliorate them.

How Some Wars Have Ended

The character of war has changed over time; unsurprisingly how they can end has also changed. No two wars are the same and none end in the same way. Nevertheless, by understanding how some wars have ended may help us to see how to end wars in the future.

The era of wars of statelets—involving the prestige and ambitions of princes—and religions was ended by the Treaties of Westphalia, establishing the modern state system. This lasted until the Revolutionary and Napoleonic Wars of the early nineteenth century, and characterized the major state rivalry of that century: Russia and the UK, Germany and France, the US and Spain, Russia and Japan.

The post-Napoleonic period ended, in fairly absolute terms, with World War I, which destroyed the imperial and monarchical system of Europe and created new states, if seething with the tensions of the old and not yet reconciled to the new. In this sense, although more technologically advanced and even more disastrous in terms of human lives and wholesale destruction, World War II was an extension of that previous great war. The wars that followed were in large part in the wake of and a consequence of the unresolved tensions of the two previous world wars, in which imperial Russia transformed into the Soviet Union, its empire extended to the Warsaw Pact and the 'peripheral' wars, of independence, nationalist assertion and ideology, reflected these unresolved tensions.

The period following the collapse of the Soviet Union and the end of the Cold War marked not an eschatological 'end of history' (Fukayama 1989: 3–18, 1992), but an upsurge in both nationalist and religious wars. Even this was a lull before our current era: marked by asymmetric warfare, cyber-warfare, specialist warfare (usually combined with the previous two) and the types of proxy wars that reflect the shifting tectonics of world power, away from liberal democratic states and toward autocracies; and their respective instrumentalist views on war, or strategic posturing ahead of possible war, as defining where the new conceptual and geographic boundaries lay.

World War II is often characterized as the archetypal war in which one side completely triumphed and the other totally vanquished. The call for 'unconditional surrender' agreed by the Allies at the 1943 Casablanca Conference was enabled by changing military fortunes which presaged a decisive shift in favor of the Allies. Military success was due to a combination of: huge manpower reserves of a global alliance; massive production capacity of the USA; geographic advantage such as, for example, the land mass of the USSR making it very difficult to conquer; and, ultimately, technology with the development of atomic weapons.

The 1939-1945 example is so prevalent in modern thought that it is sometimes assumed that total military victory is the preserve of general war between states. But this is not so. There have been many intra-state or smaller wars involving non-state actors that have also ended with total military victory, even if without an act of unconditional surrender such as there was with Germany and Japan in 1945. One example was the defeat of Chinese- and Russian-backed Popular Front for the Liberation of the Occupied Arabian Gulf (PFLOAG) in Southern Oman in 1976; another is the defeat of Tamil Tigers in Sri Lanka in 2009; a third example is the defeat by the Taliban of the Afghan government of Ashraf Ghani in 2021. In each case, war was ended through military victory not negotiated peace.

The Korean War is regarded as a war in stalemate: a perpetual conflict frozen in time. It is true that there has been neither peace treaty nor agreement on the main differences between the parties; but fighting has stopped (apart from the occasional military demonstration) and the terms of the 1953 Korean Armistice Agreement continue to hold. Over the two years previous to the Armistice, the military situation had yielded advantage for neither side despite very high levels of casualties as the Chinese, in particular, attempted to break the stalemate. The catalyst for Armistice discussions appears to have been the death of Stalin and subsequent withdrawal of Soviet technical and logistic support for North Korean and Chinese armed forces. This is an example where internal political change in one country has impacted the ending of a war in another.

In some respects, the continuing Indo-Pakistan conflict is similar to the situation in Korea, in that there has been no agreement between the parties on the main issues behind the conflict, despite four major wars and frequent outbreaks of lesser levels of armed violence. That both India and Pakistan possess nuclear weapons is probably one of the reasons why violence remains limited, keeping the situation in stasis rather than one side believing it can subjugate the other through military means.

Other forms of ending wars might be said to include the Yom Kippur War, in which military superiority of one side was clear but it did not end in complete subjugation of the other. The Iran-Iraq War ended as a consequence of neither country being able to establish meaningful military advantage over the other, in what might be termed a 'hurting stalemate'. A similar type of 'hurting stalemate' also characterized the separatist war fought by the Moro Islamic Liberation Front in the southern Philippines island of Mindanao.

In yet other cases, there has been third party intervention, sometimes by a multi-lateral body such as the UN under a Security Council resolution, for example in Kosovo or Timor-Leste; or by a military alliance such as NATO in Libya; or by a regional power such as India in Sri Lanka, or by the US and allies in Grenada; or by way of direct military support for a warring party, such as by some Western states in Ukraine. There may also be non-military interventions, such as by the EU and ASEAN in Aceh, following mediation by a non-government actor such as Crisis Management Initiative.

In each case wars have ended through absolute military victory or less, through the defeat but not subjugation of the enemy, through effectively permanent ceasefires and through negotiated agreements. Especially in the case of negotiated agreements, the role of third parties, both bilateral and multilateral, as mediators, brokers, guarantors or enforcers has been critical to the outcome. Wider understandings, too, of the end of wars being defined by more than just the absence of violence has come to play a key role in sustainably ending wars and creating a comprehensive and sustainable peace.

The book

Understanding the causes of wars, how they are fought, and how this understanding might help find their resolutions, is at the heart of this book. It is the result of a virtual conference on 'How Wars End' held under the auspices of The Australian Institute of International Affairs Victoria in Melbourne in 2022.

As with the conference, the book is divided into thematic sections, each examining a key aspect of war and its resolution. The first part of the book considers how and why wars start, which seems to be central to how and why they might be ended; followed by a typology of war, which proposes that how wars end depends on their type or character. This is then followed by a reflection on the war against the Islamic State in Iraq, using that example to examine how a war is fought and how it unfolds may change a combatant's policy objectives, including how 'winning' is defined.

The second part of the book focuses on case studies of how some wars have ended, including those in Sierra Leone, East Timor and a comparative study of the end of the wars in Aceh, Indonesia and Sri Lanka. It also includes an examination of the use of mercenaryism in Libya and concludes with two chapters looking at differing aspects of the Taliban's 2021 victory in Afghanistan. In each case, the factors that contributed, or not, to war endings are examined to create a range of experiences from which conclusions may usefully be drawn.

The third section of the book looks at an illustration of the holistic management of external conflict resolution, using the Australian experience, and concludes with an assessment of how great power conflicts may end in the nuclear age.

The next section of the book considers the questions of 'ripeness' in ending wars, along with related questions such as who is able to negotiate, with what authority and other issues that come with negotiating, along with types of interventions. The section ends with the role of UN peacekeeping/making operations and conflict resolution

The conclusion highlights that efforts to end war do not always, or even often, go according to plan and that an image of wars with neat ends is not useful in making sense of complex environments, especially those marked by a proliferation of armed non-state actors. But, that said, it posits an approximate theory for how wars end that may be helpful to understand how to terminate a conflict or, hopefully, prevent it from starting in the first place.

References

Fearon, J, 1995. 'Rationalist explanations for war', *International Organization*, vol 49, No 3, pp. 379–414.

Fukuyama, F, 1989. 'The End of History?', *The National Interest*, No 16, pp. 3–18.

Fukuyama, F, 1992. *The End of History and the Last Man*. The Free Press, New York.

Kaldor, M, 1999. *New and Old Wars: Organized Violence in a Global Era*. Polity Press, Cambridge.

Rapoport, A, 1982. 'Editor's Introduction', in Clausewitz, C. von., *On War*, Penguin Classics, London.

Wilson, W (1918). 'President Wilsonas Message to Congress', Records of the United States Senate, 8 January.

Section 1

Why War?

This section is about war: why wars happen; what form they take; and how they are fought. It is apparent that the causes of a war, its typology and how it is fought will in some measure influence how it ends; but it is not immediately obvious how. Our first three authors shed some light on this difficult area and allow us to better understand what war is, so that we are more able to see a way to its end.

Sir Hew Strachan examines the causes of war and how they have changed over time. He points out that, despite best attempts by policy-makers to draw neat arrows between policy, strategy, action and results, war very rarely meets the expectations of those who embark upon it. He remarks that it is particularly difficult to pre-determine the relationship between the causes, course and consequences of war; indeed war, once embarked upon, has its own dynamic that will cause states or other actors to alter their original objectives and, as a result, change their appetite for an end to hostilities, for better or worse.

Robert Johnson explores how wars differ from each other and develops a typology against which we can meaningfully describe individual wars. He identifies ten criteria by which wars can be characterized, including, for example, political purpose, geographical spread and intensity. Each characterization comes with its own difficulties in creating peace but, also, with its own examples of successful peacemaking that can help future peacemakers. Finally, Johnson points out that war is about power: victory results in an accumulation of power; defeat in a loss of power. Ending wars is largely about a redistribution of power.

Ambassador (and Major General Retired) Roger Noble examines the relationship between how a war is fought and how it ends. He uses the 2016 campaign against ISIS in Iraq to illustrate a number of observations, including the need to understand potential endings long before they are reached. He comments that the impact of connectivity that enables *how* you fight can now be observed by very large numbers of people globally, in almost real-time, resulting in some potentially surprising influences on the eventual outcome of the war.

DOI: 10.4324/9781003317487-2

1 The Causes Of War—And Their Consequences

Hew Strachan

'War is merely the continuation of policy by other means.' When—as they frequently do—commentators cite Carl von Clausewitz, the most profound as well as the most famous analyst of war, they do so to make this point above all others. It is as though, without Clausewitz's imprimatur, the reader will not grasp what is taken to be a central truth. It is also a warning. As Clausewitz continued, 'The political object is the goal, war is the means of reaching it, and means can never be considered in isolation from their purpose' (Clausewitz 1976: 87). Statesmen who embark on war are reminded that the means can too easily become ends in themselves, and so strip war of its purpose. On this reading, the political aim which precipitates war becomes the context within which to contain it; without policy, war's inherent violence lacks utility or justification; it can escalate to become self-fulfilling and so reverts to its primordial nature; it is simply destructive.

Liberal democracies have embraced this Clausewitzian ideal with fervour, creating a master narrative to rationalise the resort to war. It begins with policy and the expectation that it shapes strategy, which in turn determines operational and tactical choices. Strategic planning rests on the presumption that war is a linear process. Its logic is simultaneously hierarchical and sequential. In its idealised form, war is designed so that it delivers the objectives which reflect the reasons for which states have undertaken it in the first place. From its causes come its course and from its conduct emerge its consequences. According to this model a war's origins are reflected in its outcomes.

Most of the time they do no such thing. The belief that the causes, course and consequences of war trace a straight line, following a logic that is continuous, is perverse because for much of the time it is at odds with reality. Of course, there have been moments when the reasons for which states have gone to war have been fulfilled in the consequences that have resulted, but they have been rare. The dominant example in the history of modern Europe was the wars of German unification in 1866 and 1870: short, sharp conflicts which delivered on Otto von Bismarck's aim to unite Germany under Prussian domination. However, even here there is a danger of overstatement. In 1871, after the second of these wars, the chief of the Prussian general staff, Helmuth von Moltke the elder, laid down a general

DOI: 10.4324/9781003317487-3

(and regularly recycled) strategic principle—that 'no plan of operation extends with certainty beyond the first encounter with the enemy's main strength'. To drive home the point, he went on: 'Only a layman sees in the course of a campaign a consistent execution of a preconceived and highly detailed original concept pursued consistently to the end' (Hughes 1993: 45).

Hindsight hides the degree to which war is rarely the continuation of policy by other means. After it is over, the knowledge of victory shapes how the war's history is read. At its commencement, neither side, however determined on achieving success, can know its eventual outcome. Reciprocity lies at the heart of war. Each side is trying to fulfil its own aims at the expense of the other. If one prevails, the other does not. For the defeated side, the probability is that the war will be profoundly discontinuous: regimes collapse, as they did in 1815, 1918 and 1945. Even for the victors, the outcomes can be very different from those which they first intended. In 1939 France and Britain went to war to support Poland, finally re-united in 1919 after its dismemberment in 1795, and confronting—as a result of the Nazi-Soviet pact—renewed partition between Germany and Russia. France and Britain ended up on the victorious side in 1945, and Poland was reunited, but as a Soviet satrapy, not on the terms which its allies had sought six years earlier. The continuum from causes to outcomes rarely holds.

Victory can blind the victors to the contingencies of war. They read a war's outcome as evidence that their cause was right and their methods of fighting justifiable. Moltke's successes in 1866 and 1870 gave the German army an exaggerated idea of its own prowess in 1914. Despite facing much greater odds in a much bigger war, its leaders believed that operational and tactical flair could cut through strategic vulnerability. When the Cold War ended, some Americans translated the fall of the Soviet Union into a form of victory, despite the fact that it had been gained without the contingencies of actual conflict. For the United States and its allies, the successful British recovery of the Falkland Islands in 1982 and the defeat of Saddam Hussein in the first Gulf War in 1990-1991 had confirmed that war could be an effective instrument of policy. Their consequence was hubris. The wars fought since the 9/11 attacks in 2001 have revealed the challenges of maintaining any sort of continuum between causes, course and consequence. The invasions of Afghanistan in 2001 and Iraq in 2003 and the intervention in Libya in 2011 were all marked by initial operational success, but what were greeted as a quick and seemingly decisive victories transmogrified into stubbornly protracted and indecisive conflicts.

All three of these post-9/11 wars were characterised by the absence of any alignment between their declared ends and the means used to achieve them. The United States and its allies fought them as limited wars. These were not existential struggles for national survival but expeditionary interventions conducted at a distance by professional forces without full national mobilisation. However, they were not limited in their objectives, which were staggeringly open-ended. Described at their outset as the 'global war on terror', they lacked definable or achievable objectives. When in 2009 that phrase was shelved, it was replaced by another, 'the long war', which displayed a similar lack of focus.

Rather than being limited in their aims, the post-9/11 wars were limited by the means allocated to their conduct. The United States opted not to commit the full might of its military capability to their execution, and its allies followed suit. As a result, from the start the military means did not match the scale of the political aims. Because the aims were so broad and ill-defined, nobody could be quite sure what success might look like. The confusion as to the objective grew as the initial operational success gave way to insurgency and civil war. The latter meant that limited military means were no longer adequate and so the effort was increased with 'surges' in Iraq in 2006–2007 and in Afghanistan in 2010–2011. The limits were now set in terms not of effect but of time, most blatantly in 2009 when President Obama said the surge in Afghanistan was to end by 2014, regardless of whether it had achieved its objectives. However, if a war is not defined by its political objectives, it cannot simply be limited in terms of its means. The United States behaved as though it might be; at times some even seemed to believe that it would be. None of its allies questioned that assumption, opting simply to follow in America's wake. They judged their contributions in terms of scale, the size of their national contingents, more than in terms of causes or outcomes. The narrative of operational success, derived from the planning and execution of campaigns, proved continuous, but it was rewarded with persistent political failure.

All three of these post-9/11 wars came to be framed by a similar story. In the first phase, a western coalition intervened in the name of big ideals, to counter terrorism, to protect democracy and to promote good governance. In the second, when the allies confronted resistance led by terrorists and guerrillas, they responded by resuscitating the neglected precepts of counter-insurgency warfare. Over time, counter-insurgency morphed to the point where it ceased to be an operational method and became the strategy itself, a point reached in Iraq in 2006–2007 and in Afghanistan between 2009 and 2012. Now the implicit aim was to create sufficient security for an acceptable political solution to emerge without any sense of what that political solution might be. It would be shaped not by those with the military power (who were foreigners) but by those whom that military had put into office (who were seen as locals, even when they had sometimes been long-term exiles). The presumption that the latter might come to the rescue of the former neglected the political displacement created by a strong military within any polity in time of war, and particularly in a fragile or even failing state.

The effect was to separate military capability from political effect. In the final stage, at the point when the intervening coalition forces had exhausted the time allowed them by their home governments, the United States developed a new form of counter-insurgency and one whose means were even more limited. Insurgents were to be countered not by 'boots on the ground' or by 'hearts and minds', but by a combination of unmanned aerial vehicles and special forces. As with the previous version of counter-insurgency, this one too became a strategy in itself, with the situation on the ground left largely in the hands of local forces. Absent in western counsels was any real sense of what a political outcome would look like beyond the hope that it was one which the warring parties would accept as their own. One theme dominated this history: the lack of any alignment between means and ends.

Now consider an alternative narrative for Afghanistan, Iraq and Libya. Each began not as an 'intervention' but as an invasion. Coalition forces were welcomed not by subject peoples freed from the shackles of tyranny but by potential national rulers who hoped to speak for their peoples in order to seize power. That invasion was initially successful because high-intensity combat operations were still the bedrock of strategic planning and their efficient execution by well-equipped and professionally-trained armed forces had an immediate power all of its own. However, at the political level what underpinned those operations could be characterised not as counter-insurgency but as insurgency. In 2001, 2003 and 2011, when western troops intervened in Afghanistan, Iraq and Libya, they were not in the business of propping up the existing order but of toppling it. They were revolutionaries: they overthrew the Taliban, just as they removed Saddam Hussein and Gaddafi. In Afghanistan the United States refused to permit Hamid Karzai to bring the Taliban into the negotiations to establish a provisional government in December 2001 and in Iraq it conducted a purge of the government and the armed forces as complete as that of Stalin in the Soviet Union in the 1930s, albeit without the direct violence. The invasions were themselves the precipitants of the insecurity which the coalition forces had then to curb. By 2009 they were conducting what they called 'stabilisation operations' in order to end the instability for which they, at least at the local level, were in large part responsible.

The planning and execution of campaigns, because it is a process taught with consistency and system at defence academies and staff colleges in a proper professional environment, is frequently done better than the framing of the policy that shapes the decision to go to war in the first place. Moreover, both are often done without any formulation of the post-war aims that might guide both the policy and the operations. Clausewitz was wrong, at least in practice if not in theory: war is often not the continuation of policy by other means, and that is particularly so for twenty-first-century democratic states.

There are at least three reasons why that is the case. First, for democracies, the resort to war represents not a continuity but a profound change. It signifies the failure of policy, not its fulfilment by other means. In the minds of their own publics, democracies are associated with peace, order and prosperity. They don't, at least in the narrative of 'democratic peace' theory, go to war with each other and that is because they no longer regard war as a normalised element of international order.

Secondly, violence lies at the heart of war. Taking casualties and killing other people change the dynamics between powers. They make negotiation harder and compromise less possible, at least in the short term. Belligerents need to honour their dead and they do it by re-committing themselves to the cause for which their comrades and fellow citizens laid down their lives: they cannot be seen to have died in vain. That was evident in December 1916, when both German and American peace initiatives failed to end the First World War, despite the fact that by then casualties could be counted in millions. Instead of motivating a search for peace at any price, the dead became sunk costs. In 2006–2008 elements of the same phenomenon, albeit on the basis of much lower casualties, were evident when George W. Bush doubled down on the war in Iraq rather than withdrew, and in 2009, when Barack Obama found himself unable simply to end a war in Afghanistan to which he was politically but not personally committed.

Third, even if war is policy by other means, it is not a policy like other government policies. When a government adopts a policy for health or educational reform, it is operating in a less resistant environment than war. That is why policies can have continuity: they can be framed, unveiled and then implemented. If need be, they can be corrected or tweaked in the light of experience. Political debate in a parliament or national assembly may sometimes evoke the metaphors of combat but it is not war. Nobody is killed. The reciprocity in war makes war a less predictable tool of policy. The opposition is not just amending or criticising the other side's policy; it is trying to impose its own and is ready to die in the process. War therefore has an escalatory dynamic that can very quickly go in unforeseen directions, precisely because it is not under one government's control.

War therefore tends to trump policy, not policy war. The realities of the battlefield, and the capabilities at the operational level, define what is possible and necessary more than do policy and politics. In wartime, war shapes policy as much as, if not more than, policy shapes war. An example is in the pursuit of allies. Before 1914 Germany scorned the idea of an alliance with the Ottoman empire; after the outbreak of the First World War it sought it, despatching funds to the bankrupt Turks in order to force them into active hostilities against the Entente powers. By doing so, Germany hoped to draw British, French and Russian troops away from the European theatre. In the Second World War, when the Germans invaded the Soviet Union on 21 June 1941, Winston Churchill, who had said in 1937 that, if he had to choose, he might favour Nazism over Communism, saw the Russians as 'fellow human beings in distress' (Reynolds 2005: 101; Colville 2004: 351). After the war was over, Churchill would resume his struggle against Bolshevism but until 1945 he subordinated long-term goals to the immediate need for victory.

A point that we too often forget relates to this pursuit of allies in wartime. Clausewitz grew to adulthood during the French Revolutionary and Napoleonic Wars, and his ambition was to understand not their origins but their direction. His insights on the relationship between war and policy were therefore largely limited to the waging of war, in ways which have been lost by modern thinkers bent on appropriating his dictum not for what it says about the formation of strategy within war but for what it says about war's onset. On this it actually has very little to contribute. The theorists who laid the foundations of strategic thought in the late eighteenth and early nineteenth centuries were not primarily concerned with war's causes, for all that they recognised that they could determine the shape of the fighting that followed. Their interest in the role of policy in war lay more in the ways it shaped war once it had begun. When Clausewitz said war was the continuation of policy by other means he was making a comment designed to reflect the influences on a general in his conduct of the war more than on the statesman weighing up whether to abandon diplomacy for war. Moreover, Clausewitz realised that the practice of war all too often did not match the theory. He knew that, because its means are so horrific and destructive, they could trump the ends, and that war could therefore cease to be a political instrument.

On War invokes policy as an enabler of war in books I and (especially) VIII, but largely as an ill-defined abstraction. Its role is not further developed. To find out more, we have to turn to Clausewitz's prefatory note, dated 10 July 1827 and published by his widow in the introduction to the first edition of her husband's book in 1832. Here Clausewitz says that policy has two roles in war. First, it can aim at the complete overthrow of the enemy, rendering him 'politically helpless or militarily impotent, thus forcing him to sign whatever peace we please'. Secondly it can contain war by setting definable and manageable objectives. He describes these in geographical terms, and specifically as the occupation of the enemy's frontier districts. Although today we tend not to define war in terms of territorial acquisition, it is this second view of war's purpose which has largely conditioned the reading of Clausewitz's discussion of the relationship between war and policy in modern democracies. Policy is there to give war utility by making it a rational pursuit of defined and containable ends. However, in seeing *On War* and its discussion of the relationship between war and policy in these terms, we can ignore two points. First, Clausewitz goes on to say that, although his discussion of the subject moves from one type of war to the other, readers must not forget that the two are quite different and contain 'points of irreconcilability' (Clausewitz 1976: 69). Secondly, war as waged by Napoleon in Clausewitz's day more often approximated to the first type, not the second.

In other words, Clausewitz's own experience made him much more focused on war's capacity to escalate than on its susceptibility to limitation. The French Revolution set objectives which expanded war precisely because the revolution liberated the powers of the state for the purposes of war. In 1792, democratisation enabled war rather than contained it. War and policy are most in harmony when nations are fighting an existential war (as we would call it today, or 'an entire war' as Clausewitz called it). The United States and its allies may have been embarking on limited wars when they intervened in Afghanistan, Iraq and Libya, but the same did not apply to their opponents. They were engaged in a struggle to define their nations and their fight to resist a foreign invasion was unfettered and unlimited. For the insurgents, war and policy were in harmony because the objectives of both could be defined in existential and therefore congruent terms. Many of the problems encountered by the coalition forces along the political/operational fault-line, which in turn generated civil-military tensions, reflected the fact that for them there could never be such harmony within a so-called war of choice. That is why the United States and the United Kingdom so readily return to the Second World War for their models—because for them it was an existential war and therefore the last occasion that war and policy were fully in step.

Clausewitz's point was that war is not limited of itself. However, he makes a further point. War is a means to an end, not itself an end; the end is policy. That statement does not get us very far. It begs a key question because policy is itself only a means. Governments design policies to achieve objectives. In war their policy might be, for example, to acquire territory or to effect regime change or to respond to a humanitarian crisis. By pretending that policy is the aim of war, we put ourselves in a position where we can be fuzzy and uncertain as to what our specific policy objectives are.

It is worth recalling the historical context within which Clausewitz wrote. Both the principal warlords of Clausewitz's age, Frederick the Great of Prussia, under whom his father had served in the Seven Years' War, and Napoleon, against whom he himself had fought, personally united political supremacy with military command. Both had a common model: Louis XIV of France shaped the approach of Frederick the Great and, as Napoleon's powers grew as emperor, so did his assumption of the trappings of Bourbon authority (Storring 2021). The potential for unity between the reasons for going to war and the war's conduct lay in the consistency provided by autocratic rule.

Broadly speaking, Frederick got the balance right. He became an object lesson in how to do it, Napoleon in how not to. In 1740, as an heir apparent who had yet to establish his reputation as a great general, Frederick published a book called *Anti-Machiavel.* Written in conjunction with Voltaire and designed less to refute Machiavelli than to update him for the conditions of the eighteenth century, it made clear the direct link in Frederick's mind between the duties of monarchy and the use of war by the state: 'a prince fills only half of his vocation if he specializes only in the trade of war'. He amplified these points in his political testaments of 1752 and 1768 (Luvaas 1996: Chap 2). He opened his reign by fighting a war of limited territorial conquest, seizing Silesia from the Austrians. In 1756 he had to fight a major war, the Seven Years War, to hold on to what he had gained. Cause, course and consequence followed a narrative between his accession to the throne in

1740 and the end of the Seven Years War in 1763. The Seven Years War prostrated Prussia and at moments in the struggle Frederick invoked the vocabulary of existential, not limited, war, as he fought to survive. But he also laid the basis for Prussia's future power and for its acceptance in the European order of great powers. For the last twenty-three years of his life, he eschewed war for peace. His final war, that of the Bavarian succession (1778–1779), proved so limited as to be effectively bloodless. By the 1770s, his achievements shaped the military thought of the Enlightenment. Frederick and his contemporaries, it has been argued, accepted the outcome of a battle as decisive for policy, to the point of giving it legal status (Whitman 2014). Two of the Enlightenment's best-known writers, whose works influenced Clausewitz, the Comte de Guibert, author of *Essai general de tactique* (1770), and Heinrich Dietrich von Bülow, who produced *Geist des neuern Kriegssystems* in 1799, both made the links between war and policy central to their writings on strategy.

Napoleon behaved with much less circumspection in his use of war. His political reputation and the empire which he created were founded on his capacity to win victory on the battlefield. Unlike Fredrick, he could not fall back on his name and birth to legitimise his succession. His title as emperor depended on his capacity to achieve success on the battlefield. He spurned opportunities to call a halt, first in 1802 with the peace of Amiens and secondly following the treaty of Tilsit in 1807. The French had defeated the Austrians and Russians at Austerlitz in 1805 and the Prussians at Jena in 1806. By 1809, however, Napoleon was at war with Austria once again, and in 1812 he invaded Russia, launching the campaign which definitively shattered both the terms of Tilsit and the emperor's aura of invincibility. If war is the continuation of policy by other means, this one proved doubly impolitic. As he lost his wars, so he forfeited his legitimacy. To regain it when he escaped from Elba in 1815, he stressed not his authority as emperor so much as the roots of his original success—the French Revolution.

The rhetoric of the revolution spoke not just of France's 'natural frontiers' on the Rhine and the Alps but of liberty, equality and fraternity, and saw these as universal rights which others had to adopt and respect. Therefore, in 1814–1815, when the peacemakers met in Vienna to reconstruct order in international relations, they were determined to deprive war of its capacity for popular mobilisation. For them, revolution had initiated a twenty-year sequence of protracted conflict and stripped its conduct of any constraints. The European wars of 1792 to 1815 were, it seemed, fought not just between states but also between different value systems. War was the continuation by other means not simply of policy, but also of politics.

It followed that, if revolution could be prevented, so in turn could war—or at least war on this scale. The international order devised at Vienna sought to prevent both. Historians tend to forget, as they consider the evolution of armies over the course of the nineteenth century, how much their roles were shaped specifically to prevent revolution and to maintain domestic order.

As Germany and France turned to face each other once more in 1914, much of their armies' training and active duty was conditioned by the suppression of strikes, by the protection of their rulers from coups or assassinations, and by shaping the loyalties of conscripted citizens to the regimes which they served (Porch 2008: 106-12, Kitchen 1968: 115-86, Ulrich 2001: 163-8).

The distinctions drawn by contemporary writers between inter-state war and intra-state wars—civil wars, revolutions and insurgencies—are in many ways a product of the establishment by the state of a monopoly over the use of force. In the sixteenth and seventeenth centuries civil war and international war elided with each other. Domestic strife, particularly in the aftermath of the Reformation, created the opportunity for external intervention, leading to what today's theorists call 'internationalised civil war'. Political and religious loyalties were fluid, especially as monarchs struggled to assert their primacy as rulers. As they did so between the end of the Thirty Years War in 1648 and the French Revolution in 1789, states established standing armies not least to ensure that they monopolised the domestic use of violence. In doing so they furbished them as instruments to create order at home as much as to sustain their foreign policies. Although during the nineteenth century police forces replaced armies as the principal aid to the civil power, soldiers remained the final sanction of governments.

The effect was to take wars that were centrally about politics, in other words civil wars, revolutions and insurgencies, out of the mainstream of military thought. In 1848 revolutions broke out across most of Europe but they did not provoke inter-state war as the French Revolution had done in 1792. Only Russia intervened in others' affairs, doing so in Hungary in support of Austria and the monarchical principle. Thinking about strategy and operations in the late nineteenth century was conditioned not by these experiences but by the Napoleonic Wars and the Wars of German Unification, and concentrated on inter-state conflict fought by national armies. In both cases the elements within them of revolution and of insurgency, although present, were largely ignored, from the Spanish guerrillas in the Peninsular War to the French *franc-tireurs* after the defeat of the French army at Sedan in September 1870 (Förster 1987).

Pre-1914 military thought also began to eliminate the idea that states went to war for monetary gain or for territory, as Frederick the Great had done when he seized Silesia. Clausewitz's descriptions of likely objectives in war still focused on the capture of territory but from the middle of the nineteenth century international law sought not only to contain war's worst excesses but also to prevent it in the first place. War began to be characterised as unproductive and economically destructive. That would have seemed absurd to the Romans or to Napoleon. The oldest cause of war is not policy but possession: possession of others' wealth and women. Between 1815 and 1914 European states still fought wars for empire but did so against non-European peoples. Wars of conquest, although they continued, were increasingly couched in different terms.

First, states aspired to be nations made up not of subjects but of citizens. Peoples who shared the same language and culture but lived on the other side of an international frontier needed to be 'liberated' from foreign or autocratic rule. The unifications of Italy and Germany provided the obvious examples. Although, formally, both processes were completed by 1871 the outbreak of the First World War rekindled the rhetoric of unfulfilled nationhood. Italy entered the war in 1915 to fulfil the aims of the Risorgimento by extending its frontiers into Dalmatia by way of Trieste, so incorporating Italian-speakers into an expanded Italian state at the expense of south Slav peasants. Germany's war aims in the east, from the Baltic states to Austria-Hungary, reflected the ambitions of the Pan-German League. In January 1918 Woodrow Wilson's Fourteen Points included the principle of national self-determination, so igniting a rush to annex in the name of the nation-state which was legitimised by the Paris peace settlements. In developing countries outside Europe, nationalism remains a powerful cause of war, and takes other guises through ethnicity, tribalism and religion. More importantly, however, the appropriation of war as a tool to serve nationalism rekindled war's sense of purpose. The link between war and policy made war a matter of national survival.

Second, the economic drivers of war had not simply disappeared by 1914. At an economic conference in Paris in June 1916, Britain and especially France construed their blockade of the Central Powers not just as a means to win the war in hand but also as a device to suppress German competition in the war's aftermath. They saw economic warfare as a form of indirect protection which excluded Germany from world markets during the war, not only for the purpose of winning the war but also as a means to gain a head start in the race for post-war economic recovery (Soutou 1989: 233-71). The blockade back-fired. After the war was over, the fascist states responded by embracing the principles of autarchy and especially agricultural self-sufficiency. Italy and Germany went to war to secure 'living space'—Italy in East Africa in 1935 and then the Balkans, and Germany in eastern Europe in 1939 and the Soviet Union in 1941. Japan similarly expanded first into China and then turned in 1941 south to the Pacific, designating their conquests a 'greater East Asian co-prosperity sphere.'

The persistence of these expansionist and economically driven currents within war was overshadowed by the simultaneous elevation of war as a political instrument. The war of 1914 was very quickly understood as a clash of competing ideologies, as the wars of the French Revolution and Napoleon had been. Prussian militarism battled with the rule of law, autocracy with liberal democracy, a polarity confirmed in 1917 with revolution in Tsarist Russia and the entry of the United States into the war. Both sides used revolution, insurgency and subversion, the weapons of war at its most political, as instruments of inter-state war, Britain through the support of the Arab revolt against the Ottoman empire and Germany through the call to Holy War among the subject Muslims in the empires of its enemies. The first of the two revolutions in Russia kindled (and indeed was in part driven by) hopes that a liberalised Russia would more easily and completely tap the country's latent energies and material strength for the better prosecution of the war. That did not happen but the principled oratory of America's president, Woodrow Wilson, brought succour to those who had hoped for a purer progressive message than could be sustained by an alliance to an autocratic Romanov regime.

The Second World War in Europe, like the First, and not unlike the first of the wars of the French Revolution in 1792, began with a dispute over the control of national territory in eastern Europe, but soon acquired broader connotations. The launch of Operation Barbarossa on 21 June 1941 confirmed it as a war between fascism and communism. On 31 March Hitler had revised the draft instructions for the invasion of the Soviet Union, saying: 'The impending clash is more than a clash of arms; it also entails a struggle between two ideologies. To conclude this war it is not enough, given the vastness of the space, to defeat the enemy's forces. The entire territory must be dissolved into states with their own governments. Any revolution of major dimensions creates facts which can no longer be expunged. The [National] socialist idea … alone can form the domestic basis for the creation of new states and governments' (Förster 1998: 482).

Communism cut across the anti-fascist narrative constructed by the liberal democracies—and proved a more substantial and more sustained political challenge. Like democracy, communism promised a political end-state which could create a peaceful international order but proposed to get there via a very different route. In April 1917 the Germans smuggled Lenin back into Russia and, after the Bolsheviks seized power in November, many in Britain began to see them as a bigger threat than the Germans. Woodrow Wilson's declaration of American war aims in January 1918, the Fourteen Points, was a direct response to the Bolsheviks' call for the war to end without annexations and indemnities. His policies for the war were enunciated, not in advance of America's entry, not as reasons for going to war, but at a point when America had already been at war for over eight months. Moreover, their target audience was less the enemy than America's allies, fearful that their people would in their exhaustion succumb to socialist blandishments.

Although Wilson's ambitions for the League of Nations were undermined after 1919 by his own dogmatism and Congress's refusal to ratify the Paris peace settlement, Wilsonianism itself did not die. Franklin Delano Roosevelt resuscitated it in the Second World War. On 6 January 1942 he told Congress that the goal of the United States in the war, which it had finally entered a month before, was to establish four essential freedoms—freedom of speech and expression, freedom of religion, freedom from want and freedom from fear (Stoler 2007: 19). Couched in terms of universal principles, they were also expressions of America's 'manifest destiny', at once ideological in content and uncompromising in their demands. Roosevelt revived the League in a new form. It was now called the United Nations. Its members embraced values whose reach extended beyond national frontiers through a rhetoric which was sufficiently all-embracing to make them absolute. America's vision of itself, however imperfectly realised in practice, was globalised.

The Grand Alliance of the Second World War came at a price: a subsequent battle between liberals and communists for control of post-war Europe, played out in civil wars from 1943 onwards and crystallised in the Cold War. It entrenched the struggle between liberal capitalism and Communism in ways that were defined as existential. On 5 March 1946 Churchill himself declared that 'an iron curtain' had come down across the continent, from Stettin on the Baltic to Trieste on the Adriatic. In 1948 western European liberals persuaded the Americans to join them in containing communism, and outside Europe the United States led the way in fighting it in Korea and Vietnam, as well as through proxies in Africa and Latin America. The ideological differences were also used to define the objectives of insurgents in the wars of colonial withdrawal, as nationalism once again stoked the fire of opposition to foreign rule. Throughout war was presented as the continuation of policy by other means—or indeed of politics. The ambiguity between the two is well served by the German of Clausewitz, whose word *Politik* today embraces both, as *la politique* does in French.

Individual states within alliances, both during the two world wars and since in NATO and the European Union, continue to nurture individual national objectives, just as they do in other joint forums like the United Nations. The more detailed those political aims and the more they reflect regional concerns or differing cultural perspectives, the more they can become divisive within multi-lateral debates. The reverse is therefore easier: big ideas can be grandiose but imprecise, so stressing common values rather than political differences. At the same time, however, defeat becomes unconscionable as it would mean the loss of democratic rights and of individual and collective freedoms. By linking war and policy, democratic states committed themselves to a use of war which implied that it was likely to be existential.

The result is that democracies find themselves in a position which is confused: a confusion directly manifested in their determination to fight limited wars for unlimited ends. Its most important outcome, as the invasion of Ukraine on 24 February 2022 revealed, is that they are not ready to fight potentially unlimited wars defined as such in terms of both means and ends. That refusal is a logical conclusion given the collective European memory of the Second World War and the impact of nuclear weapons, but it is also illogical—a product of muddled thinking reflected in two overlapping debates—one academic conducted by international relations theorists and the other mediated in public between governments and their electorates. Neither provides a narrative sufficient to link the causes of a war to its course or to its consequences, whether a war is limited or unlimited.

Theorists of international relations are conventionally divided into two—realists, who see the state and its survival as the main drivers of international relations, and progressives of a more optimistic bent. Clausewitz appealed to the former among those who had lived through the Second World War and then the Cold War, precisely because he responded to a set of comparable experiences, when war had been similarly central to international politics. Opposed to the realists are those who argue that the destructiveness of major war should force us to find another way. The peacemakers of 1919 and 1945 responded to this aspiration first with the League of Nations and then with the United Nations, both of them multilateral institutions designed to forestall conflict through negotiation in the first instance and, if that failed, with sanctions. The roots of such thinking trace themselves back at least to the Enlightenment, with its recognition that humanity has it in its power to create a better world. Idealists believe in the influence of liberal institutions, regard international law as an ongoing project to eliminate war as well as to moderate its conduct, and embrace human rights and their promotion as core values.

To see behaviour in international relations as a dialogue between these two poles is not hard, but to see the origins of wars in terms of one or other model is. In both early 1917 and early 1941, most Americans, as they observed the wars then being fought by Germany, sympathised with the causes for which the allies were fighting, but they remained opposed to the United States entering either war. In other words, they were not ready to fight with Britain even for the principles which they held in common. Similarly, in 2013, as the civil war climaxed in Syria and the government of Bashar al-Assad released chemical weapons on its own people, the populations of most western countries expressed horror at such actions and the suffering it caused. However, that desire stopped short of military intervention. They rejected the use of war for the purposes of policy. In modern democratic states at least, most citizens are liberal humanitarians at heart but reluctant warriors in reality. Neither liberalism nor realism is independently able to explain how the governments of today's democratic states go to war. In practice apparently polarised positions converge. Even if realism may more often trigger war, liberal internationalism is frequently used to explain it once it is underway—and to justify it once it is over (Knutsen 2016: especially 185-8, 258, 308). The invasion of Afghanistan in 2001 may have been designed to eliminate a haven for international terrorism but it then morphed for many into a war to create a more democratic and just society in Afghanistan.

After 1945 the Charter of the United Nations created a set of norms which were inherently contradictory. It disallowed wars except in cases of national self-defence. It therefore followed that wars which were legitimate were 'existential' wars, or wars of national survival which by definition were likely to be major wars for those who fought them. However, as Clausewitz pointed out, the responsibility for initiating war rests not with the aggressor, but with the defender, because he decides to oppose aggression. In 2005 the United Nations added to this conceptual confusion by endorsing the Responsibility to Protect (R2P), which authorised external intervention if there were overwhelming humanitarian reasons for doing so, specifically to forestall genocide, war crimes, crimes against humanity and ethnic cleansing. The United Nations was built round the principle of national sovereignty but R2P can be used to authorise its members to reject that principle and to use war offensively, although the issue of self-defence is not at stake.

The result is that democratic states use war in ways which rest on contradictory assumptions. They begin with a presumption against the use of war by sovereign states, so reflecting the UN Charter. Governments can therefore prove reluctant to go to their national assemblies to seek approval for military action for fear it will be denied. In 1973, after the Vietnam War, the US Congress passed a war powers resolution which required the President to consult Congress before committing American troops to war. Successive presidents evaded it and Congress has abandoned the principle. In Britain, the call for a similar law grew after the invasion of Iraq in 2003. Gordon Brown, when prime minister, supported a statutory requirement for the government to seek parliamentary approval for war. In August 2013, although no such law had eventuated, David Cameron acknowledged the principle by seeking parliamentary approval for intervention in Syria. When parliament refused its approval, he abandoned the policy, although he was not constitutionally obliged to do so. No subsequent prime minister has sought to implement a war powers act. The result is that both countries can go to war without either a formal declaration of war or the approval of Congress or Parliament. They can vow to protect democracy abroad but show themselves fearful of its consequences at home.

The reluctance to face public accountability for the use of war is tantamount to an admission that war represents a failure of policy. But that underlying assumption in turn makes it harder to coordinate war and policy. Western democratic powers use war reactively, not positively, and so they find it hard to counter those national leaders, pre-eminently Vladimir Putin, who are not so inhibited. The best moment for western intervention in Syria arose in 2011 and was already passing its peak in 2013. Thereafter, Assad had the upper hand in the war but the western powers found it hard to accept the consequences, not least because it dented the 'inevitability' of eventual victory for a cause that was just, as democracy was supposed to be. Their defeat in Afghanistan in 2021 made manifest a dysfunctional approach to war's management already evident in the unfolding events in Iraq and Libya but one whose implications had not been properly digested because the west's frustrated hopes in the latter were less conspicuous than the pell-mell retreat from Kabul airport.

These difficulties are in turn glossed over by the exaggerated rhetoric conventionally used by many heads of democratic states when addressing the problem of war. The analogies with Pearl Harbor used by George W. Bush in the aftermath of the 9/11 attacks and with appeasement used by Tony Blair in the run-up to the war with Iraq in 2003 are cases in point. So too were the

United States's embrace of the 'global war on terror' or in Britain David Cameron's tendency to talk of terrorist attacks on British citizens in Tunisia and Algeria in existential terms although they presented no direct danger to the United Kingdom itself. Their successors, Barack Obama and Theresa May, spoke in more modulated and cautious tones but struggled to climb down from the priorities established by their predecessors' use of strong language. In 2022, Boris Johnson in Britain and Joe Biden in the United States (albeit with greater initial reticence) used the absolutist rhetoric of freedom and democracy when backing Ukraine in its war with Russia, while being careful not to commit their countries' armed forces to the conflict.

Through all this, over more than two decades from 2001 to 2021, the tendency has been to over-promise and under-deliver. From Bush to Biden, from Blair to Johnson, no American or British leader has proved willing to use the vocabulary of limited war in order to explain their policies, although in effect that is what has proved to be their most consistent theme. Instead, the limitations—real enough—have been manifested in the means they have used, managing their own manpower, minimising their armed forces' losses, and enabling proxies to do much of the fighting and as much as possible of the dying. Their wars were and are wars of words, where ends have once again exceeded means.

Those tendencies, to over-promise and under-deliver, are deepened by their need to act in alliance, even if only as a coalition of the willing. Alliances lend legitimacy to their actions, given the requirements of the UN Charter, and help generate sufficient combat power, given individual states' reluctance to dedicate the resources required to achieve victory and then deliver long-term stabilisation. However, by acting as an alliance, the western powers commit themselves to political purposes that tend to reflect the lowest common factor not the highest common denominator. All can subscribe to the ideas of stopping genocide or preventing war crimes, or to defending liberty or human rights; few can agree to specific war aims defined in attainable and regional objectives. In Afghanistan NATO too often prioritised its own cohesion over the stability and order of Afghanistan. Its initial responses to the Russian invasion of Ukraine in 2022 did not break that pattern.

In April 2018, the United States reacted to Syria's use of chemical weapons against its own people by executing a strike which was designed to uphold an international regime opposed to the use of chemical weapons. It avoided inflicting any loss of life and it had no further consequences for the war in Syria or its eventual outcome. That use of force goes to the heart of the problem. What good did it do? Was it even an act of war? Syria did not treat it as one by reacting in kind. It is impossible to place such an attack within a narrative of causes, course and consequence. A tactical action was used to cover a political void, not to provide a political solution.

There is a danger in reductionism. It is not wrong or unhelpful to acknowledge the role of policy or politics in causing war, but we need to separate that role from their functions in shaping war. We can too readily interpret the aims articulated so clearly during the heat of battle as reasons for going to war in the first place. In the process they become sharpened, contingency is down-played and inevitability exaggerated. Policy can and often must be the servant of war, not its master.

The problem, then, is that the intensity of a war, once it has begun, oversimplifies the narrative and so eliminates or subordinates other interpretations. The iteration of Clausewitz's 'formula' (to use Raymond Aron's title for the relationship between war and policy) begs questions that are both temporal—if policy has been the cause of war in modern Europe and even across the modern world, has it always been so?—and more fundamental. If it is generally true that war is the continuation of policy by other means, that still says very little about the proximate and particular causes of individual wars. A policy to do what? The fulfilment of a policy may be the aim of a war, but it is itself still a means to an end—to the establishment or the removal of a government, to the creation of a nation state or to the dismemberment of an empire. Don't we need to know these distinctions better to understand the causes of war? They may be second-order questions in relation to the 'formula', but they are not secondary to the decision to go to war in particular cases. States don't take up arms because they see war as the continuation of policy by other means, but because they see no other way to achieve much more specific objectives.

References

Clausewitz, C. von, 1976. *On War*. Translated by Michael Howard and Peter Paret. Princeton University Press, Princeton.

Colville, J., 2004. *The Fringes of Power: Downing Street diaries 1939-1955*. Orion, London.

Förster, J., 1998. 'Germany and the Second World War, IV: the attack on the Soviet Union.' *Militärgeschichtliches Forschungsamt*. Oxford University Press, Oxford.

Förster, S., 1987. '"Facing people's war": Moltke the elder and Germany's military options after 1871', *Journal of Strategic Studies*, vol. 10, pp 209–230.

Hughes, J., 1993 (ed). *Moltke on the Art of War: Selected Writings*. Presidium, Novato CA.

Kitchen, M., 1968. *The German officer corps 1890-1914*. Clarendon Press, Oxford.

Knutsen, T.L., 2016. *A History of International Relations Theory*. 3rd edn, Manchester University Press, Manchester.

Luvaas, J., 1996. *Frederick the Great on the Art of War*. Free Press, New York.

Porch, D., 2008. *The March to the Marne*. Cambridge University Press, Cambridge.

Reynolds, D., 2005. *In Command of History: Churchill Fighting and Writing the Second World War*. Penguin, London.

Soutou, G., 1989. *L'Or et le sang: Les buts de guerre économiques de la Première Guerre Mondiale*. Fayard, Paris.

Stoler, M., 2007. *Allies in war: Britain and America Against the Axis Power, 1941-1945*. Bloomsbury, London.

Storring, A., 2021. '"Our age": Frederick the Great, classical warfare, and the uses and abuses of military history,' *International Journal of Military History and Historiography*, 27 August 2021, pp 1–33.

Ulrich, B., Vogel, J. and Ziemann, B. (eds), 2001. *Untertan in Uniform. Militär und Militarismus im Kaiserreich 1871-1914. Quellen und Dokumente.* Fischer Taschenbuch Verlag, Frankfurt am Main.

Whitman, J., 2014. *The Verdict of Battle: the Law of Victory and the Making of Modern War.* Harvard University Press, Cambridge Mass.

2 A Typology of Wars and How They End

Rob Johnson

Do the types of war, with different pressures and dynamics, indicate how wars are sustained or how they end?

In academic and policy making circles, there are broad categories of war which are usually associated with a linear spectrum of conflict. At the lowest level, individual acts of homicide or the lethal actions of small, organized groups, while not categorized as 'war', are identified as one end of the range. The scale is invariably increased to include internal security, fighting by irregular forces, sustained civil conflicts, and organized insurgencies, with a figure of 25 combat deaths considered the threshold for measures of armed conflict (Uppsala 2021). The scale increases still further with limited wars between polities, campaigns on land, at sea, in the air and in the electronic domains, and then major wars and total wars, culminating in the prospect of nuclear, annihilationist war. While this categorisation is attractive, a linear approach neither captures the full range of types nor identifies the constituent elements that determine each type. Moreover, it offers little insight into determining how wars actually end, and whether scale affects the process.

This chapter therefore sets out ten constituent elements and shows how their relative importance creates the types that occur, not only between different wars, but also within them. Each element is given a brief evaluation before examining different disciplinary approaches. It is the interaction of these elements that not only discern a type of war, but which also gives us some insight into how a particular war might be concluded. So, first, the chapter gives some explanation of how and why a typology is constructed. Then, it addresses how a typology can assist in understanding what factors influence how wars end, with the suggestion that war is the accumulation of power, defeat is the loss of power, and peace is the redistribution of power.

The ten elements are human factors, scale, geography and resources, technologies, political purposes, organisation, intensity, duration, objectives, ways and means, and domains. Within these, we can identify the following Table 2.1:

DOI: 10.4324/9781003317487-4

Table 2.1 The constituent elements that define war

Constituent Elements	*Common Types of War (and a range of)*
I Human Factors	State to Ethnic Conflicts
II Scale	Minor to Major
III Geography and Resources	Internal, Regional, Inter-state, Internationalized, Global
IV Technologies	Technologically-led/defined to Primitive
V Political Purpose	Liberation, Revolutionary War, Separatist, Resource, Pre-emptive, Limited, Total, Annihilationist
VI Organisation	Civilianized, Irregular, Regular/State, Coalition/Alliance
VII Intensity	High to Low
VIII Duration	Short/Decisive to Protracted
IX Ways and Means	Guerrilla/Counter-Revolutionary, Proxy, Counter/Insurgency, Un/Conventional
X Domains	Cyber/Electronic, Informational-Psychological, Air, Sea, Land, Space

Typologies, a product of the scientific method of classification, are popular because demarcations and definitions are designed to aid our understanding of the world and allow us to make comparisons. The problem is that simplistic modelling does not always represent complexity accurately. To illustrate this point, in the context of war, consider the battlefield at Arras in France in 1917. From a distance, one would observe trench fighting in a set piece offensive of a defined total war. But this was also an action characterized by desperate man-to-man fighting with rifles, bayonets, spades, helmets, shovels, and grenades. It featured hand to hand encounters with men grunting, shrieking, wrestling and wielding blunt or bladed instruments (Nicholls 1990). At that level, it had much in common with a prehistoric battle. Moreover, the investment of lives over time, made concluding this particular war far more difficult than achieving a single, decisive victory.

The underlying theme of typology is periodisation. There are commonly held views about what defined each age, and thus the type of warfare. As Clausewitz noted, each age has its own theory, or grammar, of war (Clausewitz 1989: 717). Through history, the skill of manoeuvre and the ability to bring missiles and fire down upon an enemy superseded the raw face-to-face use of muscle power. Fighting in close linear formations on land and sea gradually gave way to dispersal and fighting in great depth. In 1772, the Count de Guibert noted that the scale of armies (one could add fleets) meant commanders could no longer direct formations personally (Guibert 1772). They could only be assigned objectives, locations and resources and urged to manoeuvre, to fight, and thus contribute to political goals. In the 1920s, Basil Liddell-Hart's advocacy of air and armoured technologies as permanent solutions to the killing fields of the First World War, like that of Arras, appealed to professional and public audiences eager to avoid the repetition of that costly conflict, and manoeuvre warfare developed in popularity (Liddell Hart 1929: 5). After 1945, the advent of nuclear weapons changed radically ideas about manoeuvre and the political purpose of war. They created the assumption that war was a choice between nuclear annihilation or limited conflict (Freedman 2004: 6–14). During the Cold War, a number of civil wars were exploited by the superpowers, with contingents of various size and character, from small special forces units (Borneo) to full scale deployments (Vietnam, Communist Afghanistan) as they sought to avoid direct confrontation.

In the 1990s, Mary Kaldor posited a type of war that seemed to satisfy a widespread desire to understand the post-Cold War international order, namely that wars of identity, fought by peoples affected adversely by globalisation, was the 'new' type of war that defined that period (Kaldor 1999; Kalyvas 2001: 9–100). Ending such wars proved especially problematic because of the multiple, fractured interest groups asserting their political goals, which tended to produce protraction and low intensity violence. Contemporaneously with Kaldor, Steven Pinker argued that state-on-state wars, and indeed violence in general, was in terminal decline and the world was becoming more peaceful, which appealed to those who had seen the maturing of international institutions (Pinker 2012). But state wars have continued.

The rapid advances in automation and global surveillance in the 2000s, including the use of Unmanned Air Vehicles, or drone-enabled land and air forces (in conflicts such as Nagorno-Karabagh in 2020), meant that technology re-emerged as the dominant theme of typology. The performance of Ukrainian light infantry with advanced anti-armour weapons against Russian formation reinforced the technologists' analyses. Earlier, Peter Singer, among others, captured the popular imagination with his illustration of future war fought with cyber, electronic and robotic systems, while Chris Coker has argued that, with the rapid development of artificial intelligence, this may be the last chance that humans have to determine types of war and perhaps whether humans will be able to end them at all (Singer 2009; Coker 2013: xix-xxv).

Nevertheless, while technological periodization is an appealing explanatory system, scholars of strategic culture are aware that relics of the past linger on and some characteristics, norms, and ideas are enduring, in part because of political systems, legal regimes, geography, and economic structures. It may be going too far to claim that there is a 'way of war' which would typify countries, but there are relationships between social and cultural systems, and the types of conflict that they produce (Sondhaus 2006), and thus, perhaps, the willingness to bring wars to an end. As an example, mass industrialized societies have produced very different types of war from medieval elite societies of Europe, Japan, or the Middle East. Prevailing norms may set widespread political and societal expectations and they can affect not just intellectual interpretations of war, but also the legal regimes that try to regulate and interpret it. Asserting the normative nature of peace in some developed countries has created significant challenges not only in the pursuit of policy and national defence and security, but also in defining the responsibilities and rights of state power and individual citizenship (Weber 1972; Locke 1690: ch. 7, section 87 and ch. 9, section 124). This has implications for the conduct of war, how wars are defined, and how they are concluded (Johnson 2021).

The first point is therefore the obvious one. The periodisation of war is the way that its types are usually defined. There are evident differences between wars of the ancient period and the modern. It is not only the physical features of these wars that are distinct. It is to be found in the perceived utility of war, between its desirability or necessity, set against its destructive capacity.

On the other hand, there are overlapping elements of war across periods, and some common and enduring features of war which we have usually defined as its nature. It was Thucydides who wrote that war was, ultimately, a human concern (Platias 2017). As a species, humans have sought protection or the acquisition of critical resources, to satisfy honour and credibility, and to neutralize threats. This required organisation, and a degree of sacralisation, with an understanding of its continued utility over time. Through such mechanisms, violence was transformed to a purposeful human activity, considered worthy of sacrifice. Despite the prevalence of non-state violence, the world's conflicts are still, to a large extent, decided and determined by organized polities, today in the form of state actors. In some cases, states make use of proxies, informal groups of fighters, and sometimes negotiate the control of resources, territory, or conditions. On other occasions, they orchestrate highly organized state forces. Human actors drive conflict, they resolve them, and their behaviours determine success or failure. Colonel Ardant du Picq, the French officer who authored *Battle Studies* in the nineteenth century, once wrote: 'Technological advances cannot change the human condition and therefore man's response to combat' (Picq 1946: 39). His injunction was to study the nature of the human in war, as it was the constant feature of conflict. The breaking of the determination to fight became a central aspect of thinking about how objectives could be achieved, and a war could be brought to an end.

Nevertheless, there have been variations in this human dimension of war, and this is our first constituent element to define a type of war. Political systems and types of actors wield relative degrees of power. Their importance in societies can vary, and thus their ability or willingness to wage war. There have been warrior elites, embedded within the ruling party; dynasts and tyrants with proclivities to wage war; ethnic conflicts; popular movements that have made use of war to mobilize populations; and more bureaucratized systems that have eschewed it. We should bear in mind misleading generalisations, including the myth that democracies do not wage war on each other: the idea of Kantian peace theory has been discredited by historical realities that show popular systems are just as likely to wage war and indeed are more likely to inflict casualties on a large scale (Hegre 2014: 159–172; Downes 2012: 457–489; Gat 2017) and even to avoid or delay negotiations. Our typology could therefore, with caveats, be defined by human behaviour, aspirations and organisation, including political systems.

The second element for deciding on types is that wars are often differentiated by scale, and we might hypothesize that a small war could be easier to end than a large one. Perhaps the smallest war of the modern era occurred on 27 August 1896 between Great Britain and the Sultanate of Zanzibar (Frankl 2006: 161–177). The conflict lasted between 38 and 45 minutes, involved the participation of barely 1,000 people, and was characterized by a single naval bombardment. By contrast, the Second World War engulfed the entire globe and involved tens of millions, taking almost seven years to conclude. Between these extremes are hundreds of conflicts that have seen the participation of a few thousands, to total wars that have required the service not only of millions of military personnel but also civilians in logistics, production, transportation, and agriculture in support of the war effort. To categorize war, many analysts accept 1000 or more battle deaths as the measure of what constitutes the condition (Soeters 2014). Yet the relationship of scale to conclusion is not direct. Small scale wars might be protracted and lead to very large death tolls indeed, while secondary effects of disease and famine might account for more lives.

The third element is therefore geography. Some wars are localized and contained; others sprawl across continents. The destruction of systems of trade in one region have sometimes initiated collapse in its dependencies, such as the Bronze Age collapse in the Near East in 1170BC. Other wars have either been localized or develop as regional conflicts as part of a larger struggle, such as the campaigns of North America, South Asia, the Atlantic, and Europe during the Seven Years' War (1756-63). Some wars have been characterized by a focus on a single point of some strategic significance on the earth's surface, such as a siege or the securing of a vital waterway. There are wars of separatism and wars of unification, defined by territory. Nevertheless, this category has included not just geographical space but resources and human geography too. The result can be a hybrid one of global, regional, and local issues that overlap and interact. There may be a relationship between the complexity of territorial issues, and extent, and the relative ease with which a war may be ended. However, even a war in a single, clearly bounded territory, such as Afghanistan in the late Twentieth and early twenty-first century can last decades.

Throughout history, the introduction of new technologies and techniques, our fourth element, has been heralded as transformative and has been used to differentiate wars. In terms of technologies, we create types by the materials of stone, bronze, and iron ages, by advanced metallurgy and the development of gunpowder as propellent, by the development of steam power as propulsion, then oil and chemicals which facilitated the internal combustion engine and thus armoured vehicles, aircraft, and rocketry. We further create types based on mobile telephony and radio communication, computing, synthetics, and data. Many have heralded the beginnings of a robotic, autonomous and intelligent age. Ahead lie other possibilities including neural and biological disruption, the recoding of systems and humans, and thermal warfare. The assumption has often been made, and bitterly regretted, that an advanced technological power can more rapidly overwhelm another and thus end a war quickly.

In the Western canon of literature, classical texts identified new weapons and decisive battles as the arbiter of type, while in the modern era the technological and industrial developments pervading economies and societies gave rise to evaluations that emphasized specific weapon systems, transportation, and then powered flight as the forces that gave rise to specific types of war, such as total war, or air war. In the twenty-first century, the empowerment of terrorists and insurgents, after five decades of relative stability in a global system of nation states (despite large and prolonged conflicts in Africa, Southeast Asia and the Middle East), gave rise to the judgement that a type of war is defined by these actors. Within two decades, this confident assertion was replaced by the notion that electronic communication systems, robotics and artificial intelligence would be so radically different that the essence of war would be determined by them, and that a new type of war is emerging (US Army 2017; Johnson 2014: 65–76). Yet the technological definitions could give no reliable indications of whether a war would be concluded quickly or not. As a singular factor, technologies could not account for the sheer will to fight and determined political objectives.

For political purposes, we can therefore identify intentions as our fifth element of type. Wars can be limited in objectives, escalate (or 'creep') to a wider set of goals, or manifest themselves as *guerre a l'outrance* from the outset. They can be fought for particular interests, which will influence their type profoundly. The political purpose of Caesar's unscrupulous invasions of Gaul and Britain were to establish his family's prestige in Rome, to acquire slaves, and to demonstrate the military superiority of his legions. Hitler's vast invasion of the Soviet Union in 1941 was to initiate a racial war against Slavs, create new *Lebensraum* for Germany's agricultural and industrial development, and to demonstrate the supremacy of his imagined Ayran race over so-called *Untermensch*. For Saddam Hussein, in his far more limited invasion of Iran in 1980, the purpose was to catalyse the collapse of the revolutionary regime in Tehran, acquire 'Arabistan' (Khuzestan), and consolidate his power in Iraq. Wars have been waged for colonisation (France in Algeria in the 1830s), to destroy an imminent threat (Israel in 1967), to defeat an enemy by a 'bolt from the blue' surprise (Pearl Harbor 1941), to prise open local trade (Opium Wars), as resistance to invasion (Poland 1939), as solidarity to an alliance, and to release hostages (Abyssinia 1868), and to break away from another polity (American Civil War and Biafran War). Collectively, we can group these as programmatic types, that is, those based on political purpose. This is perhaps the most useful individual factor in ascertaining how wars end, but it is not guaranteed. Two or more belligerents' intentions may be asymmetrical, or conflict so directly as to make any conclusion impossible without escalation, or some incremental or decisive edge. It may even be a question of how well organized a belligerent becomes.

The sixth element in the typology is thus organisation. The organisation of a state, its armed forces, and its systems can all have a significant effect on the types of wars that are fought and how they are concluded. Prehistoric communities appear to have waged war for the acquisition of protein, land, sites of significance, the possession of slaves, or to seize fertile women, since all were vital to their survival. Tyrannies and ancient dynasties formed societies with a privileged warrior elite and a larger, conscripted mass of levies, which determined to a large extent the types of wars that were waged. Religious motivations dominated the zealous types of war conducted by Europeans between the Twelfth and the Sixteenth Centuries. Dynastic regimes waged wars of influence and control in the eighteenth century; while the American colonists fought a democratic rebellion against Britain in the 1770s, and their desire to minimize their casualties created a protracted type of conflict. By contrast, tribal rivalries and the desire to wipe out all opposition characterized the *Mfecane* of Shaka Zulu's war in the 1830s, and similar existential drivers existed amongst Afrikaner settler resistance to the Zulus in the decades that followed. Militaristic and populist ideologies typified the wars fought by the Nazis and Soviets in the 1940s, and their uncompromising views of their enemies led to massacres and intense combat at every juncture. Such forms of war made ending them through negotiation far more difficult. The advanced nature of America's technological and informational society, and the desire to minimize casualties, was reflected in the type of warfare the United States conducted in the early 2000s, with a significant emphasis on precision strikes using guided weapons from aircraft, both manned and unmanned. Control of these operations occurred remotely, via networks of computerized data feeds, visualized in electronic imagery giving the wars a particular character. Such a type of pain-free war might lead to the view that its termination is not desirable, and there are other examples of profiting from wars' continuation.

The term 'intensity' has been used to describe types of war, and thus constitutes our seventh element. Low intensity refers to protracted conflicts where episodes of violence are episodic and even rare. High intensity is used to characterize constant operations. Yet these binary types are often unsatisfactory, for much is excluded and there is little acknowledgment of the similarities at the lowest level of a vertical scale. Conventional war, between organized, overt armed forces, is contrasted with unconventional warfare, practised by terrorists, partisans, guerrillas, saboteurs, or special forces. In reality, at ground level, distinctions are harder to ascertain. Conventional military power is sometimes deployed against insurgents. Firefights between small groups of insurgents and conventional troops can resemble a tactical encounter in a conventional war. Moreover, unconventional operations are conducted during what might otherwise be considered a conventional war. Large scale wars between organized armies have invariably produced unconventional actions and actors too. Prussian officers complained of insurgent French resistance despite conventional victories during the Franco-German War of 1870-71, and the French defeat of Spain in 1808 was the catalyst for a long guerrilla war by armed citizens and militias as well as regular forces thereafter. The sheer embitterment of belligerents may prolong war. Equally, a weaker actor may discern that a low intensity struggle, while prolonged, ensures survival; a stronger actor may be impatient to terminate a war and seek a decisive victory.

Duration, the eighth element in the typology, is closely related to notions of (un)conventional war and relative intensity. Some conflicts appear to follow the Clausewitzian ideal of short and intense actions, while others are drawn out over time, perpetuated by the resources and aspirations of the belligerents. The Hundred Years War (1337–1453) was in reality a series of pulses of fighting between England and France in the Middle Ages, but was largely conventional for the period because of the nature of the belligerents' political systems, limited resources, and their objectives, namely, to preserve the regime of governance while exchanging territory and dynastic legitimacy. In the early modern period, European wars were limited in duration by the available resources of the campaigning season, such as grass and crops for fodder and rations. In a related way, it is thought that military-fiscalism may have perpetuated wars. In India, in the late eighteenth century, the Honourable East India Company was caught by cyclical driving forces: to pay for its growing army, it required land for revenue. The acquisition of land nevertheless increased the demand for security, which necessitated a larger number of troops. The need for troops and the land to pay for them seemed to perpetuate itself. Similar assessments have been made of early modern Europe, where the duration and frequency of conflicts appeared to be connected to the systems of financing, the acquisition and retention of troops, and the costs of gunpowder armies.

Ways and means are often used to demarcate types of war, so these appear as our ninth element. There are evident differences between economic warfare, political warfare, cyber and electronic warfare, or war in space, just as there are clear variations in an air-maritime campaign, or an exclusively land war. Nevertheless, most wars in the modern era, that is, since the development of air power in circa 1900, combine ways and means. The result is that war is really a compound rather than separate elements. This is not limited only to the idea of joint operations, but refers to the inherent search for any advantage over an adversary. The Allied effort of the Second World War, for example, combined support to partisans and saboteurs, vast air-maritime campaigns in the Atlantic and Pacific, with mechanized air-land campaigns in North Africa and Europe, economic targeting in air operations, and atomic warfare. Even in wars that were not 'total', one finds examples of combinations in adroit diplomacy to build coalitions, blockades, the mobilisation of public opinion, the manoeuvre of land formations, or interdiction of trade routes. All these elements characterized the French Revolutionary and Napoleonic Wars. Another common feature of wars has been the enlistment of indigenous populations in conflicts waged by other polities, including the campaigns of the late Roman Empire, the imperial wars of Spain, Portugal and the Holy Roman Empire, the French-Indian Wars, the European colonial wars of the nineteenth century, the world wars, and recent operations in Iraq and Afghanistan. The intense debate over hybrid warfare in the early twenty-first century caused historians to assert that hybridity was nothing new and that multiple modes or domains of warfare, from biological to atomic, had been used in the past. Combinations of ways and means ultimately merit consideration in a typology in their own right. In other words, even when wars appear to 'end', they may in fact be transferred to other ways and means to continue a struggle for power.

The final element in the typology is its domains. Historically, most wars are associated with land and sea, with air making its appearance in the twentieth century (unless one includes balloon observation from the Eighteenth). These physical domains are associated with types of war, such as jungle or mountain warfare, steppe and desert wars, or urban and rural temperate warfare. At sea, there are surface and subsurface types, and perhaps littoral (amphibious) and 'blue water' categories, while air warfare may be divided into low and high altitude, and perhaps also space warfare. To these physical domains, we may add the electromagnetic spectrum, where electronic warfare takes place, and the cyberspace environment, which combines network attacks on software and various forms of information warfare. The final domain is perhaps the human cognitive one. It is here that psychological warfare is conducted, and where morale, with low or high forms of determination, is crucial. In some wars, fought in one or more domains, there may be a more certain conclusion. Land wars, for example, tend to force a decision because of occupation or the destruction of forces. However, the decapitation of an adversarial leader through a precision air strike, or the slow strangulation of a naval blockade, can affect how a war ends too.

Drawing distinctive types out of history and the present has been problematic from a scientific point of view. All war contains psychological aspects, for example, so it is a matter of emphasis when selecting examples of wars where an outcome was more clearly determined by this factor above others. Fear in the face of potential imminent destruction is common, but, collectively, some human groups can defy the threshold at which ordinarily they would seek flight, and exhibit instead the determination to fight on, even to the point of complete annihilation (Lynn 2003: 220). Self-sacrifice is remarkably common, with immense suffering for the common good, suggesting an altruistic nature at work. A variety of cohesive factors, such as discipline, training, identity, and perception of the enemy can ensure that large formations, and entire communities, can suffer immense losses and yet choose to continue resistance. Psychological resilience has enabled polities to endure surprise attacks, even overwhelming ones, and to engage in costly protracted wars, even where the chances of success have been very slim. Confidence, even instinctive overconfidence, has been a valuable element of human evolution for the conduct of war, according to Dominic Johnson (Johnson 2020). He cites George Washington's profound beliefs which sustained his faltering cause against the globe's maritime and imperial great power, Great Britain. Other apparently irrational traits have their origins deep in our species' past and were vital for survival (Coker 2021: 29–70). Nevertheless, we are not unaltered products of genetically-determined evolution (Barnett 1979: 106–125). Humans are socially conditioned too. In war, we find social restraints, instinctive reactions, and barbaric behaviours at the same time, shaped by circumstances as much as by nature. These acts can sustain resistance and therefore, by extension, a war.

There are a number of less satisfactory demarcations in the typology of war. Despite attempts to divide types of war very distinctly, the fundamentals of war have not changed over time. There is still a recognisable trinity of government, military, and the people—each with differing reactions to war—and a corresponding trinity of forces (Herberg-Rothe 2009: 204-19). Crucially, the brutality of collective violence has not changed. War remains a lethal condition, replete with enmity, passion, intensity, the determination to survive, and the desire to assert power over others. But this indicates that war is still intensely 'political', that is, it is a question of power accumulation, loss, and distribution.

Fixed types are potentially misleading. War is a protean phenomenon where diversity, developments amongst rivals, changing political objectives, and friction create uncertainties. These produce dynamics of interaction as belligerents try to gain a relative advantage over their enemies. War has often been regarded as a realm of chance, uncertainty, and of high risk. The reaction of belligerents can be varied too—some respond to a crisis with courage and stoicism, while others give way and ultimately capitulate. The contest of war has been likened to a dual, a boxing match, or a street fight, because it is characterized as a series of blows, of ebbs and flows, sometimes arousing great passions and exertions (Clausewitz 1989: Book I, Ch 1, 2; Coker 2021: 125). The asymmetry of strengths adds to the unexpected outcomes of the various collisions of war. Thus, the interactive nature of war generates significant differences even in the course of a conflict, forcing government and military planners to create contingencies and the latitude for rapid changes.

Differences, and thus types, are identified on a large scale across time and space. We have noted the traditional attribution of types of wars to stone, bronze, and iron age technologies and the subsequent eras by propellants (gunpowder, chemicals), by propulsion (steam, internal combustion), and communication systems (computing networks, data). Automation (robotics, artificial intelligence) is now the defining feature of emerging warfare. But each of these eras had transitional periods, there were often relics of the past in different regions, and different parts of the world developed along distinct lines. Some states adopted technologies, systems, and organisation from what they regarded as paradigm powers, but others adapted them more successfully for their own purposes. Yet, many failed to adapt, because of the nature of their political and social organisation, resulting in asymmetrical encounters and distinctive types of wars, such as the Anglo-Chinese Wars of the nineteenth century, and European colonial wars in Africa.

While we identify types in geographical regions that are associated with periods, such as medieval European, or Aztec, Mongol, and Ottoman warfare, there are many transitions and developments. The longevity of the Ottoman Empire defies any single categorisation, and that regime presided over the rise and fall of its Janissary corps and concluded with a modern army on the German model, before its final defeat in 1918. Moreover, war is full of unexpected developments. The interactions of belligerents create dynamics, technological development is often accelerated, and what might have begun as one type of limited conflict can escalate, or protract, into very different forms indeed. Studying a single war from the dyadic perspectives of the belligerents reveals different interpretations of a type (See, for example, Johnson 2011).

One significant problem is therefore how to incorporate changes in type, not just between wars but within them, that is to say, the demarcation and overlap between state-on-state war, civil war, wars conducted through proxies, and asymmetrical conflicts. Guerrillas can be transformed during the course of a conflict, such as the Chinese civil war of the 1940s, into 'regular' armies. Regular armies can assist and operate alongside guerrillas in the course of a campaign, such as the combination of Indian Army and Mukta Bahini insurgents in the Bangladesh War of Independence in 1971. Interventions by other states in a civil war can be transformative but may just be a phase in a much longer war, further complicating the typology. The conflict in Afghanistan that began in 1978 started as unorganized armed resistance against a communist government. It was intensified by the intervention of the Soviet Union in 1979, but their departure in 1989 caused a resumption of the civil war. While of a lower intensity by the 1990s, following Pakistan's backing of the Taliban faction, the US-led Coalition intervention in 2001 once again transformed the war. Low level terrorism gradually built into an insurgency against the international forces of ISAF and intervention forces were withdrawn in 2021. Creating typological boundaries for the Afghan conflict is therefore problematic.

The challenge of demarcating wars is manifest in time boundaries, transitions between types; whether the conflict is a grey zone confrontation or a 'hot' war, if it is of an episodic nature, or characterized by more sustained pulses; and whether it is decisive or indecisive. Some terms, in current use, add little clarity to the situation. Nomenclature and paradigms such as 'accelerated', 'hyper', 'hybrid', or 'new', applied to war are opaque and perhaps owe more to creating a policy focus than a rational contribution to the field of military science. The data used to justify them do not always support the claims. They indicate the common problems of asserting the apparent novelty of recent phenomena over historical ones. History is a useful guide to future developments, beyond the immediate horizon, because it offers enduring elements for comparison. That is not to claim there is determinism. There are, however, megacycles caused by accumulations set against relatively unchanging natural phenomena, even though, at the more detailed level, they are entirely distinct. To use an analogy from astrophysics, we recognize the phenomena of nebulae, 'black holes', and supernovae, and they recur across the universe over billions of light years, but each one is distinct, even where they follow physical patterns.

The end of the West's war in Afghanistan in August 2021 makes for a powerful case. People's war must be solved by politics and sufficient resources for the people, not the great disparities of wealth that were fostered by Western intervention. The West suffered a lack of political realism, for the conflict in Afghanistan was, at the same time, a drug war, an ideological and national struggle, and a variety of people's war. It required more than prestige building projects, the construction of a remote and inappropriate presidential political system, or the waste of millions of dollars that was syphoned away by corrupt officials. Few wanted to face up to what the war was, ignored the divided nature of Afghan politics, and claimed it was a project for 'security', democratisation, women's rights, education and health, in other words, something, as Carl von Clausewitz said, that was 'alien to its nature.'

Clausewitz recognized the chameleon nature of war, that each encounter produces some new hiatus, some confusion, and a rapid search for solutions. The novelty of each conflict, today sometimes initiated by a new technology, causes analysts and policy makers to rush to create a new discourse, doctrine, or theory. These, when examined closely, often turn out to be very familiar, and owe much to existing types of war. So, despite those who advocate that war is now obsolete and that all conflicts are essentially amongst the people, the classical theorists had already recognized the problem. Moreover, there is also a logic to the situation. When the preponderant states of the world possess such advanced means to defeat an adversary in a conventional war, it is understandable that weaker belligerents would seek instead to attack these powerful states in ways that made it far harder for them to retaliate.

In the prehistoric past, human power was defined by the combination of muscular strength, the cognitive functions of the brain, and the ability to harness emotions and instincts, to ensure survival. Humans subsequently developed the capacity to organize in social groups, using communications to create networks beyond their own family unit. The creation of tools, enhanced muscle power with synthetic strength, to create transport, weapons, armour, cantilevers, pulleys, engines, and communications systems. The combination of synthetics and networks has evolved towards artificial cognition and networks, enabling synthetic means and ways of war. Perhaps, in future war, only human emotion and instinct will be left from the original set of functions, the others being otherwise automated. We can already observe, in the battle between algorithms in the High Frequency Trading markets, that the speed of digitized communications is measured in milliseconds and the volumes of data are beyond the capability of humans to process at such a tempo. Automated weapons can track and fire at astonishing speeds, and artificial intelligence is beginning to generate entirely new strategies and synthetic combinations.

Therefore, the types of war that occur in the years ahead seem likely to feature both existing categories, but also new ones, through electronic networks, in cognitive and neural systems, and in space, not only in proximity to earth, but eventually possibly on an inter-planetary basis. On the other hand, it seems very likely that the ugly manifestations of non-state war could continue too, in parallel. We may be sure that war will continue, by one type or another. The tools we have used in the past to end war will need to be updated in each new age.

To create a holistic typology of war, we must address a range of elements, their relative importance, and their transition or interaction. The table (below) offers a list of criteria (I-X) that determine types, with a selection of examples of wars from the period after 1945. In reality, the constituent elements and types overlap.

In conclusion, we have established some fundamental elements that could be used to design a workable typology. We could identify human actors and systems, the scale of conflict, the geography and resources, the defining technologies, political purpose, organisation intensity, duration, ways, and domains. All these must be set against the context of time and space, to differentiate the forms and substance of types of war. In Table 2.2, there are some illustrative examples of conflicts since 1945, but in terms of ending wars, they give very different outcomes. Some are protracted, others short in duration; some are complex and others fought by multiple actors with separate agendas, which, while individually straight forward, complicate the ending of a war.

We have established that there are significant overlaps between the constituent elements of war. Actors and systems refer to the political basis of war, in the form of the state, non-state actors, and pseudo-state networks, and each are defined with reference to the relative power they possess, which includes geographical and cultural differences. They, and their purposes, are themselves influenced by the available resources, including those that are domestic, commercial, and the essential items for the waging of war, which would include, but would not be limited solely to, technologies. A perceived or actual lack of critical resource could catalyse and define a war, from the availability of salt to the control of the drugs trade. Resources and power

Table 2.2 Typology of War, with selected examples from the period after 1945

Elements of War	*Examples (Post-1945)*
I Human (security, critical resources, honour/credibility, threat), and human systems (calculations, systems)	Ethnic and clan conflicts: Sudan nomadic conflicts over grazing and water (and South Sudan since 2011); Oromo-Somali ethnic-territorial conflict, 2016; Nagaland conflict; Nigerian ethnic-religious conflicts; Nagorno-Karabagh conflicts of Armenia and Azerbaijan; Myanmar conflicts; Sri Lankan civil war; Central African Republic civil wars.
II Scale (including asymmetrical wars defined by relative strength or status)	Minor wars include: War of Attrition, 1967-70 between Egypt and Israel; Konfrontasi between Indonesia and Singapore, Malaya, Sarawak, and North Borneo, supported by the UK. Multinational involvement, on a large scale, in a single war include: interventions in the civil war in Syria (2012–2019) Overwhelming asymmetry: US-led invasion of Afghanistan against the Taliban, 2001 Major wars in bounded regions include the Korean War, Vietnam War, Soviet intervention in Afghanistan.
III Geographical (defined by states and spaces): Inter-state war, Civil War, Internal War, Internationalized War	Inter-state: Iran-Iraq War, 1980-88; First Gulf War, 1990-91; Invasion of Iraq, 2003–2004 Internal: Yugoslavia War, 1994 (states); Libya, 2011 (state versus non-state) Civil War: Afghan Mujahedin factions, 1989–2001 Internationalized War: Syrian War, 2011–2020
IV Technologies (Preceded by propellent and propulsion revolutions, technologies include oil-based (combustion), computing/networked, autonomous and automated	ICBMs developed in 1960s resulting in Cold War; Drone technologies (after 1960); Guided (smart) munitions (after 1990); Automated and robotic systems (such as ordnance disposal in Iraq, 2004–2009; Afghanistan, 2006-14); Satellite surveillance, navigation, timing, and tracking for precision Computing in all military calculations, from aviation to visualisation
V Programmatic (defined by purpose): liberation war, resistance/guerrilla war, revolutionary war, separatist war, war to control a local resource,	Liberation War: Bangladesh, 1971; Iraqi operations against Daesh, 2015–2019 Separatist War: Biafra, 1967; Kurdish resistance, 1970s-80s War for Resources: Afghan insurgency and drug war, 2001–2021

(*Continued*)

Table 2.2 (Continued)

Elements of War	*Examples (Post-1945)*
pre-emptive war; total war (including ethnic wars of annihilation)	Pre-emptive War: Six Day War, 1967 Total/Annihilation War: DRC, 1990s; Rwandan genocide
VI Organisation and systems, with investments in, and changes to it. The arrangement of force (such as the development of navies, armies, air forces, and rocketry; or conscription, volunteer-professionals, militias or terror cells). Groupings as state, alliance, or coalition.	The transformation of the communist elements of the Chinese Civil War from isolated militia, emphasising guerrilla operations and public political education, to a single PLA conducting conventional operations on a large scale, directed by the party elites, 1937–1949.
VII Intensity (from low intensity, protracted types, to high intensity)	Low Intensity: British police-led operations in Northern Ireland, 1972-96 High Intensity: US-led invasion of Iraq, 2003
VIII Duration (short to protracted)	Short: NATO air operations against the former Yugoslavia, 1999 Protracted: Afghanistan civil war, 1978-present
IX Methodological (defined by techniques or methods): such as proxy wars, confrontations, guerrilla war, un/conventional war, counter-/insurgency.	Naxalite Maoist insurgency in South Asia, 1967-present US-led Coalition counter-insurgency operations in Iraq, 2004–2009 ISAF counter-insurgency operations in Afghanistan, 2006–2021
X Domains	Directed cyber operations by Russia, China and North Korea against the United States and Western interests (political, financial, psychological, and infrastructural), 2008-present; attempts to conduct 'regime capture' by disinformation campaigns.

also overlap with duration, intensity, and ways or means. Domains can be classed as physical spaces, artificial networks (such as electronic communications or the electro-magnetic spectrum), but also the cognitive realm, where the purpose, intent, orchestration, limitation and escalation of war occur, but where the emotions reside too. Together, these elements establish morale, the willingness and determination to resist, accept losses, remain cohesive, and perpetuate war. As noted earlier, a war is a distinct compound of elements fused and combined.

The contexts of time and space run throughout this analysis. The latency of power, and the prospect of war, can give rise to pre-emptive attacks, while dynamic developments will shape the form and thus the type of war. The adaptation, which invariably occurs once conflict is joined, will also define the type of category in which a war is placed. Time offers a distinct marker on types of war, although conceptually types can be used for comparative purposes across different periods.

The unifying feature of the entire schema is power. We may conclude that war is a manifestation of power and the various criteria of its execution, listed in the table above, shape its general type. War is an assertion of power by one entity against another. Some wars are deflections of power or are related to a perceived slipping of power. These power relationships influence the ending of wars profoundly. Some individual elements may be studied separately in this regard and help us pose questions about the end of war: do organisations and intensity, for example, make compromising for peace or a termination of war, short of decisive victory, more difficult? In other cases, the elements must be taken as a whole. We might ask: what combinations of elements make a specific war more difficult to resolve?

Victory in war is associated with an accumulation of power, while defeat is concerned with a corresponding loss of power. Negotiations too tend to occur with reference to the relative power of the belligerents. Thus, war is a condition of contention by organized armed force, driven by power, on a large scale. Its variations are the result of the interactions of the various elements identified in this typology, set against unique contexts. Inevitably, these shape how wars end.

References

Barnett, S., 1979. 'Cooperation, Conflict, Crowding and Stress: An Essay on Method', *Interdisciplinary Science Review*, vol. 4, no. 2, pp. 106–125.

Clausewitz, C. von, 1989 edn. *On War*. Princeton University Press, Princeton, New Jersey.

Coker, C., 2013. *Warrior Geeks: How Twenty-First Century Technology is Changing the Way We Fight and Think About War*. Colombia University Press, New York.

Coker, C., 2021. *Why War?* Hurst and Co, London.

Downes, A.B. and Sechser, T.S., 2012. 'The Illusion of Democratic Credibility' *International Organization*, vol. 66, no. 3, pp. 457–489.

European Union, 2001. 'Common Council Position, 27 December'. (2001/931/CFSP), revised 2011.

Frankl, P.J.L., 2006. 'The Exile of Sayyid Khalid bin Barghash Al-BuSa'idi', *British Journal of Middle Eastern Studies*, vol. 33, no. 2.

Freedman, L., 2004. *Deterrence*. Polity, London.

Gat, A., 2017. *The Causes of War and the Spread of Peace: Will War Rebound?* Oxford University Press, Oxford.

Guibert, J., 1772. Essai général de la tactique. London: chez les libraires associés.

Hegre, H., 2014. 'Democracy and armed conflict' in *Journal of Peace Research*, vol. 51, no. 2, pp. 159–172.

Herberg-Rothe, A., 2009. 'Clausewitz's "Wonderous Trinity" as a Coordinate System of War and Violent Conflict,' *International Journal of Conflict and Violence*, Vol. 3, No. 2, pp. 204–219.

Johnson, D., 2020. *Strategic Instincts: The Adaptive Advantages of Cognitive Biases in International Politics*. Princeton University Press, Princeton, NJ.

Johnson, R., 2011. *The Afghan Way of War*. Hurst and Oxford University Press, London and New York.

Johnson, R., 2014. 'Predicting Future War', *Parameters*, vol. 44, no. 1, 65–76.

Johnson, R., Sweijs, T. and Kitzen, M., 2021. *The Conduct of War*. Routledge, London.

Kaldor, M., 1999. *New and Old wars: Organised Violence in a Global Era*. Stanford University Press, Stanford, Ca.

Kalyvas, S., 2001. '"New" and "Old" Civil Wars: A Valid Distinction?', *World Politics*, vol. 54, no. 1 (October), 99–100.

Liddell Hart, B., 1929. *Strategy*. Faber and Faber Note, London: previously published as *Decisive Wars of History*.

Locke, J., 1690. *Second Treatise of Civil Government*. London.

Lynn, J., 2003. *Battle*. Westview Press, New York.

Nicholls, J., 1990. *Cheerful Sacrifice: The Battle of Arras, 1917*. Cooper, London.

Oppenheim, L., 1952. 'International Law', vol 2, in H. Lauterpacht, (ed.), *War and Neutrality*, 7th edn., Longmans Green, London.

Picq, A. du, 1946 edn. *Battle Studies*. Military Service Publishing.

Pinker, S., 2012. *The Better Angels of Our Nature: A History of Violence and Humanity*. Penguin, London.

Platias, A.G. and Koliopoulos, C., 2017. *Thucydides on Strategy* Oxford University Press, Oxford.

Singer, P., 2009. *Wired for War: The Robotics Revolution and Conflict in the 21st Century* Penguin, New York.

Soeters, J., Shields, P.M. and Rietjens, S., eds 2014. *Routledge Handbook of Research Methods in Military Studies*. Routledge, London and New York.

Sondhaus, L., 2006. *Strategic Culture and Ways of War*. Routledge, London.

Uppsala, 2021. https://www.pcr.uu.se/research/ucdp/definitions/

US Army, 2017. 'The Operational Environment and the Changing Character of Future Warfare', U.S. Army Training and Doctrine Command, Fort Eustis, VA.

Weber, M., 1972. *Wirtschaft und Gesellschacht: Grundriss des vestehenden Soziologie*. Mohr, Tubingen.

3 How You Fight A War Matters To How It Ends: A Real-World Contemporary Case Study—Iraq 2016

Roger Noble

'War is merely the continuation of policy **with** other means'

(Clausewitz 1989: 87)

One recent author's scrutiny of the above famous quote in the original German emphasizes the word 'with' rather than the oft used 'by' as the correct English translation. This led to the following observation:

> Pursuing political objectives 'with' other means connotes adding a new implement—namely armed force—to a mix of diplomatic, economic, and informational implements rather than dropping them to pick up the sword.
>
> (Holmes 2014)

This is an important distinction. War is not purely a clinical military 'force on force' clash where the last man or woman standing wins. Rather the contest occurs in a complex context that exists in, around and endures beyond the use of force. This remains true in the twenty-first century where the 'context' is now further stretched and connected globally via the ubiquitous reality of the digital age.

How war is waged has always had a direct impact on whether and to what extent desired policy objectives are ultimately achieved. *How* you fight and the way the contest unfolds may change the very nature of the policy objectives themselves. Violence, enmity and hatred have always held tremendous power over people. How you define 'winning', how you 'win' and the parameters set around achieving a victory matter greatly and are critical to understand as you plan and conduct military action. This is as important at the national strategic level as it is through and down to the local tactical level street corner security post. In the twenty-first century everyone sees what is happening in real-time and the capacity to deceive, obscure and influence outcomes through the manipulation of information is a now continuous 24/7 reality.

DOI: 10.4324/9781003317487-5

A Case Study in How You Fight Matters for How It Ends: Counter-ISIS Campaign, Iraq 2016

Through the campaign against ISIS in 2016 in Iraq I was provided an opportunity at first hand to see and experience the critical importance of finding a way not just to win the fight but to do it in a manner that ensured strategic objectives could be secured. Through the course of a year there were multiple choices to be made that would and could shape outcomes and endings. They were made at the micro tactical level, by partners and stakeholders over which you often had limited control. These decisions are made by people of every rank and experience level across the force and by a diverse set of non-military stakeholders you can influence—*if you choose to try*. This campaign reaffirmed the challenge for modern operational art is to orchestrate tactical military action within an integrated campaign that aligns with and supports the achievement of strategic objectives, while always remaining fully cognisant of the wider complex human, political, social, physical and informational environment.

By early 2016, ISIS forces had advanced deep into Iraq holding much of the Euphrates River and Tigris River valleys. The Iraqi Security Forces supported by the International Coalition had achieved their first hard-won victories with the liberation of Tikrit and Ramadi. Notably Mosul and Fallujah remained in ISIS hands. Baghdad was subject to terrorist attacks with the enemy at the gates. Confidence in the Iraqi Security Forces which had been low was slowly growing. Victory and liberation from ISIS seemed a very long way off. By the close of 2016 most of the Euphrates River Valley would be cleared of ISIS and the attack to liberate Mosul would be well underway, spearheaded by a combined Iraqi Peshmerga force. While the fight would continue for many months, this progress exceeded all expectations.

The following three observations seek to explain why how we thought and fought in Iraq 2016 greatly influenced how it would 'end'.

> **Observation 1** Close, continuous, disciplined attention to desired strategic objectives *throughout the fighting* increases the likelihood they will remain feasible and can be achieved. This includes a focus on your own objectives and those of others including, and especially, the enemy.

Clausewitz again—'The first, the supreme, the most far-reaching act of judgment that the statesman and commander have to make is to establish … the kind of war on which they are embarking; neither mistaking it for, nor trying to turn it into, something that is alien to its nature. This is the first of all strategic questions and the most comprehensive' (Clausewitz 1982).

Strategic clarity does not necessarily readily emerge in a timely manner from the strategic level. To quote my professor Eliot Cohen—'Strategy is hard'. The variables are many, the list of unknowns is long and the relevant stakeholders are often innumerable. Core ideas and objectives are often understood but the second- and third-order outcomes and relevant parameters continually evolve and are regularly uncertain. The good and bad news is that the operational level military leaders, planners and fighters on the ground are in very real terms part of the strategic dialogue and play a key feedback/advice role on what might be achieved and under what conditions. Iraq 2016 proved to be a great example of this in action.

At one level, the Coalition objective in Iraq 2016 was very clear—defeat ISIS. These two words encapsulated the core binding element in the objectives of all the Coalition troop contributing nations. In practice, moving beyond these two words immediately got complicated.

Other stakeholders, including the Iraqis, held to this core objective but also to a range of others, arguably more critical to them, such as sustaining the Iraqi State and building the confidence of the Iraqi population through and beyond the defeat of ISIS in the field. What of the long-term challenges to be faced such as the daunting scope of rebuilding cities and infrastructure destroyed while 'winning'? Consider the complexity for the Kurdish people and their executive leadership in Northern Iraq. The defeat of ISIS was a clear interest but what of the 'post-conflict' relationship with Iraq, the future prospects for autonomy or independence and the disposition of Iraqi Security Forces? ISIS objectives and intentions, especially the central importance of the 'physical caliphate' and the link to their global branding enterprise meant that 'places'—like Ramadi, Fallujah, Mosul, Raqqa and Baghdad—were key to their ability to generate power and influence in ways well beyond their simple tactical geographic significance.

For soldiers fashioning a campaign plan, all the interests and objectives of key stakeholders needed to be understood. We rapidly learned the need to continuously engage, discuss, analyse and mesh our objectives and understand the many differences in priority, objectives and parameters around the action. We were able to distil our Coalition operational objective into two succinct words—*Get Mosul*. We learned that *how*, *when* and *in what way* we were to 'Get Mosul' would greatly impact on the timing, manner and long-term durability of 'defeating ISIS'. This required a constant dialogue with the strategic level about the objectives and parameters governing their achievement that proved iterative and evolutionary.

The key moment of Coalition self-awareness came with a recognition that the Iraqis must lead as they provided the mission essential ground manoeuvre force necessary to defeat ISIS. Winning required Iraqi infantry going through 1,000 doors under fire and Iraqi tanks advancing street by street through multiple cities and towns. It was no longer 2005; we couldn't do it ourselves if the Iraqis would not. Their plan was not necessarily our preferred plan but fighting to theirs meant we could mesh our interests and ensure the Iraqis were prepared not just to win but hold and preserve their gains. Perhaps most critically it meant they would lead—*and own*—the campaign which was central to building national and popular confidence in the Iraqi Security Forces and the Iraqi State. Our Coalition approach early on was summarized by the Land Component Commanding General (CG):

> The good news is the Iraqis have a plan. It is linear. It is sequential. It is clear. Why don't we just enable their plan and make it optimal? In a very complicated place a plan built around the Iraqi plan—it simplifies what we do, our place and contribution and ties us to the Iraqis. We are gonna kick ISIL out of Mosul. Our job is not to run Iraq.
>
> (Noble 2016)

In practice, it proved to be neither linear, simple nor clear. The enduring nature of war holds in the twenty-first century. Conflict is characterized by violence, enmity, hatred, chance, friction and uncertainty—none of which are much appreciated in National Capitals. At one point in mid-2016 the CG pointedly remarked; 'As you see the icons move you know the plan.' Yet the central idea held—build, strengthen and enable the Iraqi plan in order to achieve the Coalition objective and optimize our collective chance of sustaining the desired policy outcome—a defeated ISIS.

'Getting Mosul' was, therefore, shaped by the need to support the Iraqis and their whole nation's effort, build confidence across the population that Mosul could be liberated—and Fallujah, and Hit, and any other ISIS-held town in the Euphrates River Valley, while Baghdad is always protected. It was far more than a 'force on force' clash, as important as that is. It was a clash of ideas, brands, narratives, systems around which the fate of Iraq and the ISIS Caliphate [at least] would be decided. This needed to be done in a way that built confidence across the entire Iraqi population, including in occupied areas, which required demonstrably disciplined, controlled conduct by Iraqi and Coalition forces in combat while simultaneously building the narrative of 'inevitable defeat' for ISIS. Fighting and action were, therefore, constantly metered, resourced, sequenced, messaged and paced to hold these objectives, interests and requirements together in the face of a particularly ruthless, innovative and committed enemy. One practical example of this notion in action was the evolution of integrated 24/7 Coalition-Iraqi Strike Cells that governed, provided oversight and authorized all Coalition and Iraqi offensive kinetic strikes through, by and with Iraqi approval.

The Coalition could have *destroyed* ISIS faster—maybe. We could NOT have cleared ISIS from Iraqi cities and towns because we did not commit ground combat troops beyond advisers, specialists and supporting fires. We could certainly have used more force at different enemy targets and in a different sequence than executed. ISIS casualties and losses could have potentially been higher and faster. However, what would our relationships inside Iraq have looked like? What would have been the response of Coalition nation domestic audiences? Would even more critical infrastructure have been destroyed or more civilian lives lost? What would the confidence of Iraqis in their own system and security forces have looked like? Would the tactical defeat of ISIS have been sustainable if a 'gap' in presence, resources and confidence emerged? What if the diverse key stakeholders who cooperated to achieve victory across Iraq did not have the time or opportunity to cooperate to the depth and manner that eventuated? What about Syria and the simultaneous actions to secure Raqqa? How you do it matters greatly to how it will end.

> **Observation 2** How you organise and conduct the fight matters and must include a focus on understanding likely endings as they emerge—long before you get there.

Popular culture and interest are often focused on the tactical fight. The Iraq 2016 experience reinforced the central importance of being able to win in close combat: the most testing of environments. Training, equipping, leading, organising, supporting and enabling this effort was rightly a central line of effort in the Coalition campaign and was critical to winning. ISIS adopted tactics that required the Iraqi Security Forces to close into direct fire range in complex urban environments. Failure to succeed at this tactical task would have meant defeat or stalemate for the Iraqis and, by extension, for the Coalition.

What is less obvious but equally important is recognising the critical importance of the organisation for campaigning and integrating into the complex operating environment. An attack to liberate a city like Mosul or Fallujah is a national whole of society effort in which the vital interests of many agencies and individuals are directly engaged. An advanced, primarily western, military Coalition is arguably one of the best trained, equipped and resourced entities to plan, think through and execute complex actions. But it needs to be focused on doing so; not just destroying the enemy.

Critically the military—Coalition and Iraqi—needed to see and understand the wider context and engage and shape well outside any military chain of command—if we were to achieve our own directed ISIS-focused strategic end-state. The final plan to liberate Mosul was ultimately Iraqi by design and was driven by a multitude of critical requirements many of which had nothing to do with the preferences of the attacking tactical military force. Adopting an approach predicated on integration, support, information sharing and unity of effort means looking far beyond the simple and important 'force ratio' calculations so beloved of Staff Colleges.

In the twenty-first century, this demands an 'all domain' approach—land, sea, air, space and information/cyber. There is a pressing need to build and reinforce narratives that are often set by other actors. The expertise and understanding of the environment (physical, human, informational) requires a system tuned to collection and understanding beyond 'Red Force' laydown and major combat systems. Only this style of approach enables the constantly metered, resourced, sequenced, messaged and paced actions that are so essential to achieving policy objectives.

Without such an informed and integrated approach, it is more than possible to 'defeat the enemy' in a narrow tactical sense but lose the war as a result of failing to understand the implications for the wider context. For example, striking an ISIS legitimate military target engaging Iraqi forces during the battle for Fallujah from inside a water treatment plant was both lawful and necessary from the perspective of friendly tactical forces. However, the possible destruction of the water treatment plant would have significantly increased the risk to the civilian population from cholera, a key risk identified by the United Nations and Iraqi Authorities, so the approval to strike from the air was denied in the interests of the post-combat stabilisation objective and support for the liberated civil population. Fighting in a manner that goes beyond what *can* be done to what *should* be done is key to shaping post-conflict endings.

The quality of campaign planning and analysis impacts the likely outcome. This is certainly not new, but it remains true in the twenty-first century. Adherence to the traditional military campaign planning doctrinal tenets, adapted to the contemporary environment, leads to a forward-focused, analytical approach that enables the timely identification of critical opportunities, gaps, risks and issues. Using and thinking through traditional concepts such as centres of gravity, campaign end state and objectives, decisive points, operation objectives, effects, lines of operation, sequencing, phasing, operational reach, culmination, risk and contingencies allows for a systematic, disciplined consideration of multiple possible futures and ultimately leads to a better understanding of how to shape to the preferred future. A brilliant campaign planning culture cannot guarantee success, but its absence greatly reduces your chances of achieving your objectives.

Perhaps most important is commander's visualisation—the ability to think through, understand and explain the practical path to success. The experience of Iraq 2016 reaffirmed the importance of this essential skill and adopting systematic campaign approach. It revealed critical capabilities early such as the availability of limited Iraqi military bridging assets or the sequence and strength of forces required through time and space to advance, clear and then hold major population centres. It revealed critical targetable enemy vulnerabilities such as their communications system, fissures in the ISIS force makeup and the exposure of key leaders. Vital, if seemingly mundane, logistic requirements were also starkly revealed such as tank readiness and tank ammunition availability, the effectiveness of internally displaced persons management and support and access to many thousands of protective concrete T-walls. Such an approach was key to reducing friction, exploiting opportunity and served to grow a comprehensive understanding of the complex operational context.

In execution, there must be a relentless focus on understanding, learning and adapting in order to achieve objectives and shape outcomes. There are no 'set and forget' plans. Contemporary war requires modern military (and civil) organizations to prioritize evaluation and assessment. A constant review of assumptions and the achievement of milestones requires a systematic collection and analysis of relevant data and inputs. You should not 'mark your own homework' and you need to be prepared to change your plan based on evolving understanding. We learned your understanding will constantly evolve if you are looking. Awareness of confirmation bias and command self-interest is key due to the most human of risks outlined below:

> ... experts neutralize dissonant data and preserve confidence in their prior assessments by resorting to a complex battery of belief-system defences that, epistemologically defensible or not, make learning from history a slow process and defections from theoretical camps a rarity.
>
> (Tetlock 1999: 335–366)

You must seek to measure what matters, not what is easy and possible to measure. Diversity of perspective and input is a strength. On arrival in Iraq many of our performance and effectiveness measures focused on friendly force actions—bombs dropped, sorties flown, leaflets distributed, messages sent. Overtime a fundamental shift to assessing intangibles such as popular confidence in the Iraqi Security Forces—essential to mission success—became increasingly important. This required an 'all source fusion' mindset that harnessed expertise and resources well beyond the traditional military areas of expertise. This approach generates a genuine 'learning organisation' that tracks performance, identifies opportunities and gaps and enables critical understanding about where the whole effort is headed.

Organisations too narrowly focused or believing their own uncontested narrative of success place policy objective realisation at increased risk. We saw 'real-time' the relationship between Iraqi battlefield success and the confidence of the Iraqi people leading to this driving our understanding of the sustainable rate of advance and the best sequence for resourcing priorities.

> Fact be virtuous, or vicious, as Fortune pleaseth.
>
> Thomas Hobbes, *Leviathan*

> **Observation 3** How you fight (or how you are seen to fight) is more important than ever because people are watching, all the time, everywhere—and many of them decide and influence the outcome.

War is fought by humans against humans. It has been accurately characterized as both a free creative act and a clash of wills. The cognitive battle has always been central to war and strength of will is arguably the ultimate arbiter of victory. War is distinguished by physical violence and danger, but it is also a clash of ideas and beliefs whose outcome is ultimately dependent on the resolve of the participants.

In 1651, Thomas Hobbes understood the utility and impact of 'facts' in the contest of wills. The influence of a fact was dependent on the circumstance in which it surfaced. As of 2022, it is more difficult than ever to distinguish fact from fiction and the reach of information is now global, universal and immediate. Consider the running dialogue around Ukrainian resistance and Russian Army morale supported by facts, images, stories, tweets, rumour, claims and complete falsehoods. In this sense all modern wars are now 'counter-insurgency like' where a comprehensive narrative and effective communication with multiple audiences is central to 'deciding who wins.'

In 2016 in Iraq the narrative contest between ISIS and its combined adversaries was of paramount importance inside Iraq and extended well out into a global network of audiences. ISIS was characterized by a capacity to exploit the information environment to build their brand, reinforce their core narrative, gain recruits and resources and undermine their enemies. They were masters of using tactical activity to reinforce their strategic messaging across multiple audiences. The images of men in black waving black flags and advancing in pickup trucks lasted in the media and consciousness for many years after those very same trucks would not dare to venture into the open.

Since war is a contest, the enemy's view on the outcome matters. In the final days of the successful attack to liberate Fallujah in 2016 members of the senior Iraqi National Security team eagerly pointed out what they saw as a fundamental shift in the ISIS narrative that indicated the enemy understood they were losing. On 26 May 2016 they pointed to new messaging embedded in an ISIS speech by spokesman Shaykh Abu-Muhammad al Adani al Shami:

> Do you believe that defeat means losing a city or a land? Were we declared defeated when we lost cities in Iraq and dwelled in the desert without (control of) a city or a land? Will we be declared losers and you victors should you take Mosul, Surt, Ar Raqqah or all the cities and returned to how we were in the first instance? Definitely not. Defeat is to lose the will and the desire to fight.
>
> (Noble 2016)

As the Iraqis well understood, this was the first formal recognition and public admission—*by the enemy*—that they would not hold their 'caliphate'. Until this point the physical caliphate—holding ground—had been fundamental to the ISIS narrative and legitimacy. Close attention to the endless, turbulent information environment identified this as one key indicator of an enemy recognition that they believed they needed to begin to shift their own strategic objectives and narrative in response to Coalition and Iraqi tactical success and sophisticated, credible messaging.

The example above highlights one of the least obvious and most important aspects of the campaign in Iraq 2016. The significant effort, led by the Iraqis, to develop a coherent, resilient national narrative while simultaneously contesting ISIS core messaging was a key contributor to an ultimately successful outcome. This clash of credibility was central to confidence, morale and effectiveness on both sides. The Iraqi plan and sequencing of operations was as much shaped by narrative objectives as any tactical physical objectives. Ground manoeuvre was conducted in ways to achieve information-specific objectives and to reinforce messaging. The narrative around the wars ending was at least in part built, nurtured and grown by fighting in a way that both proved and communicated the outcome.

Therefore, the effort to liberate cities was both a physical military action but also a central component in messaging to multiple audiences. Therefore, *how* cities were liberated, versus cleared, mattered greatly. The attack to liberate Fallujah was purpose designed to build a narrative of success which included a tactically high-risk scheme of ground manoeuvre designed to rapidly secure key points in the city in order to claim 'liberation'. It was not designed to kill or clear all the enemy from the city, street by street. Remarkably, a role and place were found for *all* key stakeholders, including the Iraqi Army, Iraqi Police, Counter Terrorism Service, the Popular Mobilisation Forces, the Coalition and the Iranians. Civil society and international organisations were actively engaged and integrated into the liberation effort. This jigsaw of stakeholders and tasks factored in the many interests and constraints to be found across this diverse set of situational friends and 'frenemies'. All were united to defeat ISIS and support Iraq. It was notable that, after the successful seizure of key terrain and declaration of 'liberation,' many ISIS fighters fled the city in vehicle columns that were then subsequently engaged by Coalition fires well outside the city proper. Success in Fallujah was then turned back on ISIS and used to reinforce the Iraqi narrative of 'inevitable victory'. Considerable effort was extended by the Coalition to improve the Iraqi Security Forces capacity to effectively message and communicate, for example, through the training and equipping of satellite-enabled combat camera teams and the dropping of millions of information leaflets (which can't be jammed!).

Our Coalition evaluation and assessment indicated success in how the Iraqis and Coalition fought led directly to further success. Tactical success bred confidence, sustained hope in occupied areas and relentlessly undermined ISIS's own internal narrative around the ever-diminishing 'physical' caliphate. Underneath the narrative messaging was a grinding, deadly close fight where risks to combatants were extreme and repetitive. The Iraqis regularly emphasized the breadth of the internal Iraqi coalition bound together in the common cause of defeating ISIS. The not easily organized participation and full support of the Iraqi Kurds in the liberation of Mosul is one clear example of the power and influence derived by such partnerships that are then transmitted to the widest audiences. It was clear that how the war was fought was key to determining how it would end and evolve post-conflict—both enabling the collapse and tactical defeat of ISIS and generating enough confidence in the Iraqi Government and the credibility of alternate possible futures.

A quick look at 2022 begs the question as to whether perception now trumps reality. How you are *seen* to fight is now at least as important as how you are actually fighting. The Iraq experience of 2016 confirmed that the truth still has power—when proven and communicated—it goes direct to credibility. What is clear is that more people in more settings can now track and see the war *as it goes* whether they are watching actual reality or not. Those concerned with how war ends must acknowledge this enduring, chaotic, digitally supercharged reality.

Conclusion: Make Your Own Endings

How wars are fought directly shapes and determines what can be achieved and how wars 'end,' or at least how the post-conflict world will continue. The nature of war is enduring with the common recurring characteristic that it is, first and foremost, a clash between people. Every war occurs in a specific context and the participants seek to achieve policy objectives using force alongside a diverse range of other means. In the twenty-first century the conduct of war is now immediately highly visible globally; enabled and amplified by the ubiquitous digital information environment.

The campaign in Iraq 2016 reinforces the conclusion of the need to recognize how you fight shapes the ultimate achievement of policy objectives. Participants, particularly operational military leaders, need to understand this and adopt a method and approach that defines, refines and relentlessly targets the achievement of the strategic objectives within the context of the specific conflict. The feedback and advice between strategists and tacticians matter. The campaign in Iraq was a complex and adaptive balancing act that resulted in the dissolution of the so-called ISIS Caliphate and an end to any 'state-like' ISIS governance and organisation. ISIS didn't end, nor were its ideas and aspirations eternally extinguished. However, post-major combat operations ISIS remnants were faced with a functioning Iraqi state, protected by an organized and tested security force, operating within an international, domestic, multi-agency, whole of society context that had a proven record of resilience. Perfect—no, but effective—yes. While Coalition forces remain in Iraq their scale and scope are much reduced, and their purpose has now evolved well beyond the combat mission focus of 2016. None of this was a given or even looked likely in 2015; it was earned and built by the effort and sacrifices of many. The complexity of the real-world means that how you fight, adapt and shape the environment will significantly impact what can be achieved and under what conditions.

Douglass McArthur is reported to have said: 'The best luck of all is the luck you make for yourself'. The ending of wars should be viewed along similar lines. The endings of wars can, at least in part, be made, shaped and influenced by how you fight them. It is certainly true that other factors may well limit any participant's capacity to shape or influence an eventual outcome. It is, however, a soldier's duty to fight tirelessly and systematically with the knowledge that outcomes are rarely the result of pure chance or 'other factors' but are often fundamentally made and built by those active in the arena.

References

Clausewitz, C. von, 1989. *On War*, Ed. and trans. Howard, M. and Paret, P. Princeton University Press, Princeton.

Holmes, J., 2014. 'Everything You Know About Clausewitz Is Wrong: A botched translation of Clausewitz has had an enduring impact on our thinking on warfare.', *The Diplomat*, 12 November.

Noble, R., 2016. Operational Notes and Records, unpublished.

Tetlock, P. 1999. 'Theory-Driven Reasoning about Plausible Pasts and Probable Futures in World Politics: Are We Prisoners of our Preconceptions?' *American Journal of Political Science*, vol. 43, no. 2, pp. 335–366.

Section 2
Case Studies

No two wars are exactly the same; and no two wars end in exactly the same way. In Section 1, Robert Johnson explored the multiplicity of types of war and developed a number of characterizations, but even within a single characterization there is seemingly a bewildering variety of war. But that does not mean that useful lessons cannot be learned from previous experience which can be applied to our understanding of current and future wars. The trick is knowing which lessons are applicable and which are not: to discern this we need sufficiently deep understanding of individual past wars, to reflect on what happened and why and in what circumstances: what worked and why; and what didn't work and why. Only in this way can we begin to reach a level of understanding that might help identify what may be applicable in any particular current or future war.

This section focuses on a number of in-depth case studies, each of which has sufficient richness to enable reflection on different aspects of how wars end. Otherwise, as the apocryphal general is always doomed to re-fight the last war, so the peace-maker is always doomed to try to recreate the last peace.

Our first case study is from Libya. Alia Brahimi studies the use of mercenaries on both sides of the Libyan conflict and concludes that they have greatly impeded efforts to negotiate peace. For those who employ mercenaries to do the fighting and dying, the cost of continuing the conflict is less than if one's own people are bearing the burden; the incentive for a negotiated outcome is therefore much reduced. Further, in Libya's case, the majority of mercenaries have been funded by foreign states, all of whom have a stake in the outcome of the conflict as it verges on proxy war, greatly complicating efforts within Libya to negotiate a peaceful outcome between its warring parties.

The East Timor case study by Clinton Fernandes demonstrates that success in war does not always come through force of arms. Despite overwhelming Indonesian military strength relative to East Timorese resistance, East Timor eventually gained its independence through a mix of international pressure and changes in Indonesian internal politics. It was different in Sierra Leone: Richard Iron explains that, there, the rebels had to be physically defeated before they were prepared to negotiate a meaningful

DOI: 10.4324/9781003317487-6

peace. Previously, rebels had exploited peace attempts purely to permit space and time to regain military strength. He also examines the utility of concessions to an organization that has no popular legitimacy bar that which comes through the power of a gun.

In many ways, the wars in Aceh and Sri Lanka were fought for similar reasons and took place largely simultaneously. But the outcomes, both triggered by the 2004 Indian Ocean tsunami, were very different: Aceh negotiated peace with Indonesia, but Sri Lankan authorities conducted a full military campaign to eliminate Tamil opposition. Professor Damien Kingsbury compares the circumstances of why these two wars ended in such different ways.

The final case study is the fall of the Republic of Afghanistan, covered in two chapters by David Kilcullen and William Maley. Kilcullen explains how and why the Afghan military collapsed so precipitously, while Maley discusses how the USA framed the conflict incorrectly, not understanding that the Taliban was, to a large extent, a proxy of Pakistan and, if the USA wanted to negotiate, it should have been speaking with Pakistan, not the Taliban. It also raises the issue of 'not losing' as a legitimate war aim: sufficient success does not necessarily require absolute victory. Both authors highlight the incompetence of US diplomacy over several years that led to the 2021 debacle.

4 Libya's Mercenaries Crisis: 'I Am Here To Protect The King Of Kings'[1]

Alia Brahimi

On 1 July 2019, a rigid hulled inflatable boat docked in Valetta, Malta, after a 36-hour sea crossing from Benghazi in eastern Libya. On board were twenty guns-for-hire from multiple countries, including South Africa, the United States, the United Kingdom and Australia. The decision to abort their mission and evacuate from Benghazi had been a swift one. The aim of Operation Opus was to supply a warlord in eastern Libya with US-made aircraft, intelligence support and an assassination squad. But the customer, Khalifa Haftar, head of the self-styled Libyan National Army, had become angry at the sub-standard kit that was delivered. According to the United Nations, it was Haftar's displeasure that prompted the mercenaries to flee (UN 2021a: 31).

The abortive operation cost $80 million, with planning, procurement and logistics activities across eight countries. The financial management was run through three UAE-based companies. At least 31 individuals from 6 nations were involved including Erik Prince, the founder of the private security firm Blackwater which was linked to the notorious massacre of Iraqi civilians during the US-led occupation in 2007. The FBI later stated that it would investigate Prince's role in the plot, and in particular the attempt to market a modified crop duster as a military aircraft for Haftar (Cole 2021).

The spectacular bungling of Operation Opus belies a larger trend in the Libyan civil war, in which tens of thousands of mercenaries have been more successfully ensconced. One of the darker realities of the conflict is that Libyans have been killed and their country ravaged by foreign soldiers of fortune.

Over the last decade, the war has been powered by layers of actors from outside the contested territory. Both rival blocs—an internationally-recognized government in Tripoli and the self-styled Libyan National Army (LNA) based in the east—have recruited from sub-Saharan Africa. The LNA, in particular, has addressed its manpower shortage through thousands of mercenaries from Chad and Sudan (UN 2019a: 10), including feared militia groups from Darfur who, according to the United Nations, were 'seeking to increase their presence in Libya in pursuit of profit' (UN 2018: 9).

DOI: 10.4324/9781003317487-7

More significantly, state-level actors have made little effort to conceal their enthusiasm for a mercenary-led model of proxy warfare in Libya, undermining regional stability and endangering an important international norm.

'Nobody is Fooled'

It is an open secret that one of the main levers of support from Russia to Haftar's LNA is a two-thousand-strong mercenary contingent from the Wagner Group, a Kremlin-backed private security company (Khalel 2020).

As a private fighting force, the Wagner Group cut its eye teeth in eastern Ukraine, supporting Russian-backed separatists in 2014 and conferring its brand with ultra-nationalist credentials. Its founder, Dimitry Utkin, has been described as a Nazi sympathiser: a veteran of Russian military intelligence, it is reported that his callsign was 'Wagner', in honour of the German composer so beloved of, and appropriated by, Hitler (Rácz 2020). In Syria, thousands of Wagner military contractors augmented the beleaguered president, Bashar Al-Assad's, forces since 2015, offering the Kremlin a way to conceal the human cost of its adventurism, as well as a facility to push boundaries without having to answer for human rights abuses. They also directly engaged US commandoes in a four-hour firefight in February 2018, when a column of pro-Assad forces advanced towards a US-controlled area near Deir Ezzour (while the US sustained no casualties in the battle, roughly 100 of the Wagner Group's men were killed). Individuals and entities linked to the Wagner Group are known to be rooted in various sectors of the Syrian economy, from electricity to hydrocarbons, based on such arrangements as the promise of a 25% share from any oil and gas production at fields they helped to 'liberate' from Assad's opponents (Sukhankin 2019).

Wagner Group personnel began training and equipping LNA forces around 2015, but in October 2018 the relationship markedly expanded. The month that the Wagner Group accelerated the transfer of its troops to Libya, its head Yevgeny Prigozhin was caught on camera at an official meeting in Moscow between Russian Defence Minister Sergey Shoigu and Haftar (Novaya Gazeta 2018). Wagner Group contractors went on to engage in an assortment of activities in support of Haftar's LNA, from armed combat to disinformation campaigns, vehicle repair, target spotting, sniping, directing artillery fire and booby-trapping civilian areas. They flew MiG-29 fighter jets (AFRICOM 2020a) and controlled an expansive military airbase at Jufra. As US State Department official Thomas D. Smitham (2022) summarised, the Wagner Group's purpose was 'to promote instability for profit.'

It's worth noting that since 2015, the Wagner Group has been active south of Libya, in fragile states such as Sudan, Madagascar and Mozambique. In the name of training local troops, the Kremlin-supported company has battled anti-government forces and muzzled dissent. In the Central African

Republic, a contingent of roughly 1,500 Wagner Group mercenaries has helped to prop up and coup-proof President Faustin Archange Touadéra's government in exchange for gold and mining concessions (Bax 2021). Close on the heels of the announcement by France that it would withdraw its roughly 2,500 soldiers from Mali, where they had sought to quash a jihadist threat since 2013, Wagner Group troops began constructing a military base near to the airport in Bamako in December 2021 (Doxsee 2022). As Raphael Parens (2022: 5) described it, a staple of the Wagner playbook is for it to become involved with a country's military, and then launch a direct relationship between the target country and Russia's military, with a view to establishing a new client state. Referring to the Kremlin's deep involvement in Wagner Group activity, former French Foreign Minister Jean-Yves Le Drian observed that 'even if Russia denies it, nobody is fooled' (AFP 2021a).

On the state level, the UAE is accused by the US Department of Defense of being a major financier of the Wagner footprint in Libya (US DOD 2020: 37; Philp 2020). This is alongside equipping and paying pro-LNA mercenary groups from Darfur, which UAE officers seemed to directly coordinate (UN 2021b: 23).

For its part, and on the other 'side' of the civil war, Turkey transported 5,000 Syrians to fight for the Tripoli-based forces in Libya (IISS 2021), reportedly paying them $600 per month (SOHR 2020). As with those from sub-Saharan Africa, many of the fighters from war-ravaged Syria are both combat-hardened and otherwise destitute. They were transferred to Libya to compose the manpower for two opposing, broadly ideological configurations. The more revolutionary, ostensibly Sunni or Islamist fighters who challenged the repressive regime of Bashar Al-Assad were flown in by Turkey to support the Tripoli-based government; the counter-revolutionary camp led by Russia, the UAE and Egypt took the lead in sourcing fighters, including pro-Assad Syrians, for the avowedly anti-Islamist authoritarian, Khalifa Haftar.

This foreign mercenary build-up helped to enmesh the prospects for peace in Libya within a wider geopolitical web, involving the global competition between NATO and Russia, the regional battle between autocracy and Islamist-leaning democracy, and a major gas dispute between nearby countries in the eastern Mediterranean.[2]

As such, the mercenary build-up in Libya holds significant implications for the international order.

Impacting International Order

Libyan dictator Muammar Gaddafi had long employed Tuareg fighters from Mali and Niger as the backbone of his elite security corps and so-called Islamic Legion. When faced with the unprecedented revolt against his 41-year rule in 2011, he relied heavily on mercenaries from sub-Saharan Africa in his repression, bestowing them with identity cards that stated: 'I am here to protect the King of Kings' (Chulov 2011). After his downfall, many of these heavily armed and well-trained combatants returned to Mali, reigniting a Tuareg separatist rebellion. In 2012, they seized large tracts of territory in the north and declared an independent state. However, almost immediately Tuareg gains were violently usurped by jihadist factions (Brahimi 2013). This tipped the north of the country into war, prompting a large-scale military intervention from France in 2013, and the further destabilisation of the Sahel region.

Similarly, the Chadian rebels who killed President Idriss Déby on 20 April 2021 were based in Libya, where they amassed money, arms and battlefield experience as guns-for-hire. From 2019, they fought alongside Haftar's LNA (UN 2021a: 83), garrisoning with Wagner Group forces at the Jufra air base and earning access to advanced weaponry, some of which was allegedly supplied by the UAE (Walsh 2021). As a ceasefire came into effect in Libya in October 2020, groups of these fighters returned home, touching off another round of rebellion that was quickly energized by the slaying of Déby.

As Russia met stiff resistance to its invasion of Ukraine in February 2022, losing thousands of troops and seven generals in the opening month, up to 1,000 Wagner Group mercenaries were relocated to eastern Ukraine (Anilkumar 2022), bringing with them 'artillery, air defences and radar that the group was using in Libya' (Schmitt 2022). In response, Turkey closed its air space to Russian military flights originating in Libya and Syria (Al-Atrush 2022). Furthermore, both Syrian and Libyan nationals were hired to fight in Ukraine, with Haftar reportedly agreeing 'to repay President Putin by sending his loyalists' (Ball 2022). This evolving matrix involved not only the introduction of Libyan fighters into the mercenary pool but also the transplantation of Syrian mercenaries with frontline experience in their home country and in Libya, into a third warzone. On the Wagner Group side, its Russian veterans saw combat in Ukraine, followed by Syria, Libya, and then a return to the Donbass region in eastern Ukraine.

While these experiences demonstrate the radioactive spread of violence through mercenaries, the casual escalation of mercenary use in Libya also points to the weakening of a once robust international norm.

Weakening A Norm

As Sarah Percy has demonstrated, mercenary use waned after the mid-nineteenth century and the practice was considered profoundly objectionable from the 1960s onwards. In fact, by the late 1990s, the norm against mercenary use 'was so powerful it has become puritanical' (Percy 2007b: 218). The prohibition emerged from the idea that they lie beyond the control of the state, thereby posing 'a practical threat to states and a threat to the normative idea that states ought to have a monopoly on the use of force'. Furthermore, mercenaries are driven by financial motivations and therefore 'do not fight for a cause deemed by society to be appropriate' (Percy 2007a: 371). Beginning in the 1960s, there were more than 100 UN General Assembly resolutions condemning mercenaries, and where funding for mercenaries may have existed, that support was covert: 'there was no attempt to hire mercenaries openly' (Percy 2007a: 373).

The normalisation of mercenary use in Libya takes place in the context of major, state-level foreign interference. In 2019, the UN Envoy to Libya, Ghassan Salame, argued that external investment in Libya risked surpassing national involvement (United Nations 2019b). Despite a UN arms embargo, Turkey, Russia and the UAE between them transferred to Libya fighter aircraft, battle tanks, armed drones, military advisers, professional soldiers, disinformation capabilities, anti-aircraft artillery systems and surface-to-air missile batteries (Pack 2020: 5)—and they control their own military bases.

Leaving to one side the question of mercenaries, the troubling moral implications of this level of outside support have been laid bare in numerous incidences of young Libyans being killed on home soil by extraterritorial parties. In one incident in January 2020, 26 unarmed Libyan cadets were killed at a military academy in Tripoli by a Chinese Blue Arrow missile fired from a Wing Loong II drone in support of Haftar and likely operated by the UAE (BBC 2020).

The UN called repeatedly for foreign powers to end their interference in Libya, to little effect, and an October 2020 ceasefire agreement included a clause on the withdrawal from Libya of all mercenaries and foreign fighters within three months. The deadline was effectively ignored for more than one year, with an estimated 20,000 mercenaries and foreign fighters remaining, and plans for some of their facilities to expand.[3] When flickers of progress on the mercenary file began to glow towards the end of 2021, these were quickly snuffed out by the major political crisis catalysed by the cancellation of the December 2021 elections and the subsequent self-declaration of yet another parallel government. As a member of the 5+5 Joint Military Commission tasked with agreeing the process to remove mercenaries from Libya, Brigadier General Al-Fitouri Gribel stated in March 2022 that 'if the conflict inside Libya appeases and a parliament, a president, and a unified government are formed, we might have a chance to expel the mercenaries, even if by force' (Zaher 2022). Given the ambitiousness of these preconditions—which essentially represent a solution to the Libyan crisis—a significant mercenary contingent will likely remain in Libya for the foreseeable future. While thousands of hired guns will continue to redeploy to Ukraine, particularly those associated with the Wagner Group and other Kremlin-linked private military companies, the maintenance of a mercenary presence close to southern Europe becomes a valuable device for President Putin as hostility with NATO intensifies.

One survivor of a 2018 suicide bombing in Tripoli described how he had come face-to-face with one of the attackers moments before the explosive vest was detonated. In his account, he fixated on the fact that the assailant was clearly not Libyan (author conversation, Tunis, October 2018). The foreign origins of the bomber seemed to imbue the event with an added layer of moral violence. As James Pattison (2014a; 2014b) describes it, being a mercenary and using mercenaries is seen as one of the worst moral failings in conflict. Mercenaries themselves sometimes recognize this wrongdoing. One Sudanese commander explained in 2019: 'I know that we are mercenaries and we are not fighting with honour and dignity ... but this is temporary, we will go back home after we are done with our mission here' (Burke 2019). This imagined homecoming entails returning to Sudan to fight the post-Omar al-Bashir transitional government—illustrating the intersection between private armies as moral failings and their direct impact on international stability.

It is worth considering whether a different moral status ought to be accorded to mercenaries when they are deployed to defend a legitimate government, as opposed to attacking it. For example, the South Africa-based private military company hired by the government in Sierra Leone in 1995 is widely recognised as an important factor in defeating rebels from the Revolutionary United Front and restoring democratic rule (Spearin 2009: 1097). The Turkish government seemed to make a similar argument when its foreign minister criticised those who 'suggest ... the Turkish presence in Libya is equivalent to that of illegitimate groups' (AFP 2021b). Turkish officials argued that its foreign fighter contingent was in Libya at the invitation of the UN-backed government in Tripoli and could not be compared to the Wagner Group's support to the eastern rebel and warlord Khalifa Haftar. According to Turkish security sources, 'considering the two sides ... equal is not right' (Aslan 2021).

Syrian mercenaries themselves eschewed the comparison, invoking as morally relevant the cause for which they were fighting: 'we are not mercenaries ... we were invited by the Libyan army and the Libyan people and we oppose dictatorship' (Wehrey 2020). A just cause, involving the defence of democracy against tyranny, may well alter the moral balance (though it is notable that the archetypal successful intervention in Sierra Leone, as well as the Libyan case, involve resource-rich nations, which refocuses attention on intentionality and motivation). However, in the final analysis, the maintenance of a government which cannot defend itself without recourse to a private army may not be worth the strong legitimising effects of a state's open employment of mercenary fighters.

Mercenaries and Ending Wars

The widespread use of mercenaries in Libya's current war has likely played an integral role in several key calculations.

In the first place, the decision taken by Haftar in April 2019 to invade Tripoli was probably unimaginable without recourse to hired guns. Not only was forced conscription of Libyan nationals a highly sensitive issue within the eastern areas controlled by Haftar's LNA, but the reality is that the foreign mercenaries fighting on his behalf were likely financed by third parties (US DOD 2020: 37; Rondeaux 2021: 11). Furthermore, in his siege of the capital, Tripoli, mercenaries were critical to Haftar's combat capability:

> The addition of tighter coordination, anti-drone capability, expert snipers, and advanced equipment allowed the Libyan National Army to make small yet consistent advances into the capital's suburbs. Thus, over the autumn of 2019, Russian fighters became an essential component of Haftar's operation.
>
> (Harchaoui 2021)

The mercenary factor also influenced the duration of hostilities, forestalling an early LNA capitulation when its forces became bogged down outside the capital; and indeed, the Syrian fighters who joined pro-government militias on the frontlines were probably an important factor in that successful defensive effort. Access to non-Libyan fighters may have also stretched the timeline in other ways: for example, it could be easier for the leaders of a military campaign to tolerate foreign deaths for psychological as well as political reasons. By the same token, it may be harder to reach the point of combat fatigue with fungible private soldiers in reserve. What is certain, however, is that alongside the third-country mercenaries who were imported into the Libyan battlespace were foreign geopolitical dynamics which prolonged the conflict and raised the stakes.

Mercenary use may also have a bearing on the way in which combatants conduct themselves in relation to international humanitarian law, with emerging testimony from Libya implicating mercenaries in grave human rights abuses (OHCHR 2020). This would mirror the multiple reports of Wagner Group war crimes in the Central African Republic, including torture, rape and summary executions (OHCHR 2021). In theory, private military entrepreneurs could approach warfare more clinically, given the absence of personal investment in the strong emotions surrounding a conflict; however, on the other hand, mercenaryism may be more conducive to demonisation and violations of the laws of war, on account of the marked absence of kinship bonds and the fraying of accountability mechanisms. Once a full reckoning has been made, it is more likely that the Libyan experience will demonstrate a tragic link between mercenaries and abuses, particularly on the part of the Wagner Group, given its established track record in sub-Saharan Africa, Syria and Ukraine.

In terms of ending the war, there are several elements brought out by the Libyan case. Firstly, the mercenary dimension created an early point of failure in the October 2020 ceasefire agreement—it was mandated that all 20,000 foreign fighters would depart Libya within three months of its signing, yet none were withdrawn even one year later—which undermined both momentum for peace and the credibility of the deal itself. Secondly, and consequently, the peace effort was overly preoccupied with the mercenary issue, thus overwhelming local conflict drivers with regional ones. Thirdly, the push to achieve a ceasefire was complicated by reckless, escalatory activity on the part of mercenary groups—for example, the laying of landmines and improvised explosive devices around Tripoli by the Wagner Group (AFRICOM 2020b). Finally, with thousands of allied proxy fighters paused but not repatriated, neither side was fully incentivised to seek national reconciliation and to promote a comprehensive solution to the Libyan crisis.

The use of mercenaries leads to the marketization of conflict in a way that fundamentally destabilizes some of our most deep-seated ideas about the ethics of war, including what cause it is legitimate to fight for, under whose authority, and with what motivation. It also frequently sustains itself through the exploitation of desperate and vulnerable young men, as demonstrated by the thousands of Syrians, Chadians and Sudanese fighters on both sides of the frontline in Libya, including boys under the age of eighteen (OHCHR 2020). But, beyond the moral realm, the entrenchment in Libya of the Wagner Group, and other Kremlin-linked private military contractors, under NATO's southern flank is a reminder that mercenaries can represent an important geostrategic vector. In the Libyan case, this reality aggravated an already intractable conflict and continues to re-shape it in new and adverse ways.

Notes

1 This chapter originated in an article for 'MENA Source', an online platform of the Atlantic Council.

2 A longstanding disagreement over the demarcation of maritime boundaries and 'exclusive economic zones' in the Mediterranean – and therefore offshore hydrocarbons exploration and drilling rights – involves Turkey, Greece, Cyprus, Israel, Libya and Egypt. It was intensified by the signing of a memorandum of understanding on maritime sovereignty between Turkey and Libya's UN-backed government at the time, the Government of National Accord (GNA). The agreement was signed in November 2019 at the same time as a security pact that deepened Turkey's investment in the Libyan conflict on the side of the GNA, which proved decisive.

3 Turkey appeared only to entrench its mercenary presence at the Watiya air base west of Tripoli, and reportedly planned to expand it eastwards to the Misrata naval base.

References

AFP, 2021a. 'UN Urges CAR to Cut Ties with Wagner's Mercenaries Over Rights Abuses'. *France 24*. Agence-France Presse (Online) 28th Oct. Available at: https://www.france24.com/en/africa/20211028-un-experts-urge-car-to-cut-ties-with-russia-s-wagner-group-over-grave-accusations-of-rights-abuses (accessed April 2022).

AFP, 2021b. 'Libya Urges Turkey "Cooperation" Over Foreign Troops Pullout'. *France 24*. Agence-France Presse (Online) 3rd May. Available at: https://www.france24.com/en/live-news/20210503-libya-urges-turkey-cooperation-over-foreign-troops-pullout (accessed April 2022).

AFRICOM, 2020a. 'New Evidence of Russian Aircraft Active in Libyan Airspace'. *United States Africa Command.* 18th June. Available at: https://www.africom.mil/pressrelease/32941/new-evidence-of-russian-aircraft-active-in-li (accessed October 2021).

AFRICOM, 2020b. 'Russia, Wagner Group Complicating Ceasefire Efforts'. *United States Africa Command.* 15th July. Available at: https://www.africom.mil/pressrelease/33008/russia-wagner-group-complicating-libyan-cease (accessed October 2021).

Al-Atrush, S., 2022. 'Russia Reduces Number of Syrian and Wagner Troops in Libya'. *Financial Times.* 28th April.

Anilkumar, R., 2022. 'British Intelligence Says Russia's Wagner Group Deployed to Eastern Ukraine'. *Reuters.* 28th March.

Aslan, D., 2021. 'Turkish Forces, Wagner Group in Libya Cannot be Considered Equal'. Daily Sabah (Turkey). (Online) 18th May. Available at: https://www.dailysabah.com/world/africa/turkish-forces-wagner-group-in-libya-cannot-be-considered-equal (accessed May 2021).

Ball, T., 2022. 'The World's Soldiers of Fortune Find a New Paymaster: Putin', *The Times* (London), 24th March.

Bax, P., 2021. 'Russian Influence in the Central African Republic'. *International Crisis Group.* (Online) 3rd Dec. Available at: https://www.crisisgroup.org/africa/central-africa/central-african-republic/russias-influence-central-african-republic (accessed March 2022).

BBC, 2020. 'UAE Implicated in Drone Strike in Libya'. (Online) 28th Aug. Available at: https://www.bbc.co.uk/news/world-africa-53917791.

Brahimi, A., 2013. 'Jihadi Opportunism'. *Al-Majalla Magazine.* 5th February.

Burke, J. and Salih, Z., 2019. 'Mercenaries Flock to Libya Raising Fears of Prolonged War'. *The Guardian.* 24th Dec.

Chulov, M. and Smith, D., 2011. 'Libya: Gaddafi's Army of Mercenaries Face Backlash'. *The Guardian.* 2nd Sept.

Cole, M., 2021. 'Project Opus: Erik Prince and the Failed Plot to Arm a CIA Asset-Turned-Warlord in Libya'. *The Intercept.* (Online) 26th Feb. Available at: https://theintercept.com/2021/02/26/erik-prince-jordan-libya-weapons-opus/ (accessed April 2022).

Doxsee, C., Thompson, J. and Bermudez, Jr. J., 2022. 'Tracking the Arrival of Russia's Wagner Group in Mali', *Centre for Strategic and International Studies.* (Online) 2nd Feb. Available at: https://www.csis.org/analysis/tracking-arrival-russias-wagner-group-mali (accessed February 2022).

Harchaoui, J., 2021. 'The Pendulum: How Russia Sways its Way to More Influence in Libya'. War on the Rocks (Online) 7th Jan. Available at: https://warontherocks.com/2021/01/the-pendulum-how-russia-sways-its-way-to-more-influence-in-libya/ (accessed February 2022).

IISS, 2021. 'The Prospects for Libya's Interim Government', *International Institute for Strategic Studies, Strategic Comments*, vol. 27, no. 6, pp. iv–vi. April.

Khalel, S., 2020. 'US Says 2,000 Russia-backed Mercenaries are Fighting in Libya'. *Middle East Eye* (Online) 17th June. Available at: https://www.middleeasteye.net/news/us-africom-2000-russia-backed-wagner-mercenaries-libya-khalifa-haftar (accessed April 2022).

Novaya Gazeta, 2018. *Yevgeny Prigozhin in Talks with the Libyan Field Marshall.* (Online video) 9th Nov. Available at: https://www.youtube.com/watch?v=kU9TgMTniww&t=37s (accessed April 2022).

OHCHR, 2020. 'Libya: Violations Related to Mercenary Activities Must be Investigated'. *United Nations Office of the High Commissioner for Human Rights* (Online) 17th June. Available at: https://www.ohchr.org/en/news/2020/06/libya-violations-related-mercenary-activities-must-be-investigated-un-experts (accessed March 2022).

OHCHR, 2021. 'CAR: Russia Wagner Group Harassing and Intimidating Civilians'. *United Nations Office of the High Commissioner for Human Rights* (Online) 27th Oct. Available at: https://www.ohchr.org/en/press-releases/2021/11/car-russian-wagner-group-harassing-and-intimidating-civilians-un-experts (accessed March 2022).

Pack, J. and Pusztai, W., 2020. 'Turning the Tide: How Turkey Won the War for Tripoli'. *Middle East Institute.*

Parens, R., 2022. 'The Wagner Group's Playbook in Africa: Mali'. *Foreign Policy Research Institute.* (Online) 18th March. Available at: https://www.fpri.org/article/2022/03/the-wagner-groups-playbook-in-africa-mali/ (accessed March 2022).

Pattison, J., 2014a. *The Morality of Private War: The Challenge of Private Military and Security Companies.* Oxford University Press, Oxford.

Pattison, J., 2014b. 'The Ethics of a Mercenary'. *Oxford University Press.* (Online) 2nd Nov. Available at: https://blog.oup.com/2014/11/mercenary-ethics-ukraine/ (accessed April 2022).

Percy, S., 2007a. 'Mercenaries: Strong Norm, Weak Law'. *International Organisation*, vol. 61/2, pp. 367–397.

Percy, S., 2007b. *Mercenaries: The History of a Norm in International Relations.* Oxford University Press, Oxford.

Philp, C., 2020. 'UAE Accused of Funding Russian Rebels in Libya', The Times, London, 3 December.

Rácz, A., 2020. 'Band of Brothers: The Wagner Group and the Russian State'. *Centre for Strategic and International Studies.* 21st Sept. Available at: https://www.csis.org/blogs/post-soviet-post/band-brothers-wagner-group-and-russian-state (accessed February 2022).

Rondeaux, C., Imhof, O. and Margolin, J., 2021. 'The Abu Dhabi Express: Analyzing the Wagner Group's Libya Logistics Pipeline & Operations'. New America Foundation. 3 November 2021.

Schmitt, E., 2022. 'More Russian Mercenaries Deploying to Ukraine to Take on Greater Role in War', New York Times, 25th March.

Smitham, T., 2022. 'Opening Remarks'. *North Africa's Transatlantic Relations Amidst Change and Continuity.* Atlantic Council, Washington DC. 24th Feb.

SOHR, 2020. 'Turkey Reduces Salaries of Mercenaries from $2000 to $600'. *Syrian Observatory of Human Rights* (Online) 9th Sept. Available at: https://www.syriahr.com/en/183498/ (accessed March 2022).

Spearin, C., 2009. 'Back to the Future? International Private Security Companies in Darfur and the Limits of the Executive Outcomes Example', *International Journal*, vol. 64/4, pp. 1095–1107.

Sukhankin, S., 2019. 'Russian PMCs in the Syrian Civil War: From Slavonic Corps to Wagner Group and Beyond'. *The Jamestown Foundation.* 18th Dec. Available at:

https://jamestown.org/program/russian-pmcs-in-the-syrian-civil-war-from-slavonic-corps-to-wagner-group-and-beyond/#_edn69 (accessed December 2019).

UN, 2018. 'Letter dated 5 September 2018 from the Panel of Experts on Libya Established pursuant to Resolution 1973 (2011) addressed to the President of the Security Council'. *United Nations Panel of Experts on Libya*. Available at: https://digitallibrary.un.org/record/1640692?ln=en (accessed October 2021).

UN, 2019a. 'Letter dated 29 November 2019 from the Panel of Experts on Libya established pursuant to resolution 1973 (2011) addressed to the President of the Security Council'. *United Nations Panel of Experts on Libya*. Available at: https://digitallibrary.un.org/record/3838591?ln=en (accessed October 2021).

UN, 2019b. 'Foreign Involvement in Libya Must be Stopped, Top Official Tells Security Council, Describing Race Against Time to Reach Peaceful Solution, Spare Lives'. United Nations Meetings Coverage SC/14023. 18th Nov. Available at https://www.un.org/press/en/2019/sc14023.doc.htm (accessed October 2021).

UN, 2021a. 'Letter dated 8 March 2021 from the Panel of Experts on Libya established pursuant to resolution 1973 (2011) addressed to the President of the Security Council'. *United Nations Panel of Experts on Libya*. Available at: https://documents-dds-ny.un.org/doc/UNDOC/GEN/N21/037/72/PDF/N2103772.pdf?OpenElement (accessed March 2021).

UN, 2021b. 'Letter dated 13 January 2021 from the Panel of Experts on the Sudan addressed to the President of the Security Council'. *United Nations Panel of Experts on the Sudan*. Available at: https://documents-dds-ny.un.org/doc/UNDOC/GEN/N20/382/45/PDF/N2038245.pdf?OpenElement (accessed November 2021).

US DOD, 2020. 'Lead Inspector General for East Africa And North And West Africa Counterterrorism Operations I Quarterly Report to the United States Congress July 1, 2020 - September 30, 2020'. *United States Department of Defense Office of Inspector General*. Available at: https://media.defense.gov/2020/Dec/04/2002546287/-1/-1/1/LEAD%20IG%20EAST%20AFRICA%20AND%20NORTH%20AND%20WEST%20AFRICA%20COUNTERTERRORISM%20OPERATIONS.PDF (accessed March 2022).

Walsh, D., 2021. 'Where Did Chad Rebels Prepare for Their Own War? In Libya'. New York Times. 22nd April.

Wehrey, F., 2020. 'Among the Syrian Militiamen of Turkey's Intervention in Libya'. *New York Review of Books*. 23rd Jan.

Zaher, J., 2022. 'Al-Murtaziqa Fi Libya: Maseer Marhoon Bi'listiqrar wa Al-Intikhabaat'. *Al-Sharq Al-Awsat*. 7th March. Available at: https://aawsat.com/home/article/3515461/%C2%AB%D8%A7%D9%84%D9%85%D8%B1%D8%AA%D8%B2%D9%82%D8%A9%C2%BB-%D9%81%D9%8A-%D9%84%D9%8A%D8%A8%D9%8A%D8%A7-%D9%85%D8%B5%D9%8A%D8%B1-%D9%85%D8%B1%D9%87%D9%88%D9%86-%D8%A8%D8%A7%D9%84%D8%A7%D8%B3%D8%AA%D9%82%D8%B1%D8%A7%D8%B1-%D9%88%D8%A7%D9%84%D8%A7%D9%86%D8%AA%D8%AE%D8%A7%D8%A8%D8%A7%D8%AA (accessed March 2022).

5 Indonesia's War Against East Timor: How It Ended[1]

Clinton Fernandes

Indonesia invaded East Timor in December 1975 after more than a year of political and military destabilization. It annexed the territory the next year and occupied it until 1999. It received military and diplomatic support from a number of countries, the most important of which were the United States and Australia. This article shows that Indonesia's control rested on three pillars: its military superiority over the East Timorese resistance; the support of the international community; and its determination to retain the territory. This chapter focuses on the termination of the conflict. It begins by discussing Indonesia's invasion, the East Timorese resistance, international solidarity and the actions of the international community from 1975 till 1998. It builds on this discussion to explain how Indonesia was forced to withdraw in 1999. Indonesia retained its military superiority until the very end but lost both international support and its will to retain the territory as a result of East Timorese resistance and international solidarity.

Indonesian Interest in East Timor

Indonesia had not shown any interest in annexing the colony known as Portuguese Timor during its own struggle for independence from the Netherlands.[2] Its officials 'categorically abjured' any claim to the territory after independence (Weatherbee 1966: 683–695, 690). When indigenous political organizations emerged in East Timor after the Carnation Revolution in Portugal in April 1974, the Indonesian establishment was not opposed to the prospect of independence. In June 1974, Indonesia's foreign minister provided written assurances that supported Portuguese Timor's right to self-determination. However, the popularity of an indigenous organization known as the Revolutionary Front for an Independent East Timor (Frente Revolucionária de Timor Leste Independente—FRETILIN) led to Indonesian hostility. FRETILIN called for decolonization, land reform, administrative reform, popular education and the development of small industries based on primary products like coffee (Hill 1980: 9). Australian intelligence reported that FRETILIN's leaders

DOI: 10.4324/9781003317487-8

> … persistently stressed the need to develop a political system best suited to the economic and social environment and to an independent East Timor. They seemed to have been deeply committed to the development of cooperatives in commerce and agriculture as a means of improving the living standards and economic power of the indigenous Timorese. Nevertheless, they insisted that free-enterprise arrangements for Chinese entrepreneurs and foreign business interests would continue indefinitely.
> (Hill 1980: 9)

The reasons for Indonesia's hostility to FRETILIN are not hard to understand. The Suharto regime was opposed to grassroots political mobilization in Indonesia. Its view was that 'people in the villages' should not 'spend their valuable time and energy in the political struggles of parties and groups' but should 'be occupied wholly with development efforts' (Murtopo 2003: 45–46). FRETILIN's energetic commitment to working in the villages and its pursuit of land reform and public education flew in the face of this doctrine. This is not to say that FRETILIN was a model of libertarian political thought; rather, its work in mobilizing East Timorese villagers would have created a popular democratic alternative within the Indonesian archipelago. This was intolerable to the Indonesian regime, as regional policymakers acknowledged. Michael Curtin, the head of the Indonesia section at Australia's Department of Foreign Affairs, wrote that FRETILIN was 'the sort of party we would have welcomed, even encouraged, anywhere else than in Timor … If an independent and politically radicalized East Timor were to make a go of it, with political and economic help not to Indonesia's liking, it would certainly become something for discontented Indonesians to look to' (NAA 1974-5).

Given the geopolitical context of the Cold War, Indonesia described FRETILIN as communists. The reality, as Australian intelligence observed, was that most of its leaders 'were practising Catholics; of the ten main FRETILIN leaders, at least four attended mass daily.' It was 'a socialist-oriented party, but few members of the Central Committee seemed familiar with Marxist philosophy.' Its key leaders 'persistently stressed the need to develop a political system best suited to the economic and social environment and to an independent East Timor' (NAA 1978). However, Indonesia's President Suharto had come to power by eliminating the Indonesian Communist Party and crushing a strong peasant movement that had supported it. He created a more permissive environment for international financial institutions and business corporations. Suharto therefore earned the support of the United States and Australia. This support persisted during the Cold War years and beyond.

Australian officials were also aware of the presence of oil and gas in the Timor Sea. An independent East Timor would have asserted its own rights to these resources. In September 1974, Australia's Prime Minister Gough Whitlam advised President Suharto that he was 'in favour of incorporation' of East Timor by Indonesia although 'obeisance has to be made to self-determination' (NAA 1974). Indonesia launched Operation Komodo, a campaign of covert destabilization against East Timor. According to an advisor to President Suharto, the 'original idea was to train locals … to subvert and then fight FRETILIN, and through that process to take over.' But FRETILIN had too much support, he wrote, and 'our miscalculations, in hindsight, were obvious' (Wanandi 2012: 205). Operation Komodo was unsuccessful. Indonesia then launched an armed but still covert action known as Operation Flamboyan. After more than a year of steadily increasing pressure, it launched a full-scale military invasion known as Operation Seroja on 7 December 1975.

The Invasion of East Timor

Indonesia enjoyed enormous military superiority over the East Timorese resistance from the very beginning. With a strength of 423,000 Regulars (Army 230,000, Navy 40,000, Air Force 33,000 and Police 120,000), the Indonesian Armed Forces were, on paper at least, one of the largest militaries in the region (NAA 1977). By comparison, the East Timorese resistance had no air power, no armoured assets, and the only artillery of note were four 75-mm artillery pieces. Around 15,000 personnel had obtained reservist training from the Portuguese military, although not all of them were in the resistance. Within two months, all major population centres were in Indonesian hands. Indonesia annexed the territory in July 1976 (Hoadley 1977: 133). The resistance withdrew to the mountainous, thickly forested hinterland, accompanied by many tens of thousands of civilians. A military stalemate took hold by December 1976. FRETILIN could not dislodge Indonesia but neither could Indonesia prevent FRETILIN from organizing a functioning society in the mountains, or providing enough food crops and basic health care to the civilians who flocked to its lines.

From August 1977 onwards, Indonesia used OV-10F Bronco aircraft, obtained from the United States, to target agricultural areas and other food sources such as livestock with conventional munitions and with Soviet-supplied napalm, known as Opalm. The Bronco could be operated from the most rudimentary airfields, and its slow flying speed allowed Indonesia to attack villages more effectively. United States support turned the tide in favour of Indonesia. Indonesian forces that invaded East Timor 'were armed roughly 90%' with US equipment (US Congress 1977b: 59–64). When it ran out of inventory, the US helped it replenish its arsenal and supplied it with A-4 Skyhawk ground-attack aircraft (Nevins 2005: 53). According to declassified Australian intelligence reports, Indonesia received 'the greater part of her military aid from the US, and the remainder from Australia' (NAA 1976a). It is thus false to assert, as some writers have done, that 'Australia looked the other way when Indonesia was preparing to invade,' or that 'the United States looked away' (Power 2002: 146-7, see also Peake 2012: 73–83).

Both countries aided and abetted Indonesia's invasion and occupation of East Timor.

Indonesia's military operations caused food shortages and illnesses, forcing the East Timorese to come down from the mountains and surrender. The surrendering population was detained in camps that were ill-suited to their welfare. No medical care was available. Often there were no toilets and the only shelter was under trees. Indonesia prevented detainees from growing food. It gave them a small amount of food on arrival, often after extorting family heirlooms, jewellery, traditional beads or sexual favours (CAVR 2005: 1346). Diseases such as cholera, diarrhoea and tuberculosis ensured that most people who were sick died. A detailed demographic study later concluded that East Timor's death toll as a result of the occupation was approximately 204,000 or 31% of the population (Staveteig 2007: 14). This is the largest death toll relative to the total population since the Holocaust. The overwhelming majority of the deaths occurred during the famine, whose most deadly phase occurred during a nineteen-month period in 1978 and 1979.

Indonesia extinguished independent political activity at the village level. It also executed key FRETILIN leaders, many of whom had surrendered after being promised amnesties (CAVR 2005: 229). It preferred execution rather than imprisonment to prevent the re-emergence of the resistance. On 26 March 1979, it declared that East Timor had been pacified. Accordingly, it established Sub-regional Military Command 164 (Korem 164), which was subordinated to the Regional Military Command (Kodam), headquartered in Bali. Korem 164 was based in the capital of Dili. It oversaw thirteen Military District Commands (Kodim), which in turn oversaw 62 Military Sub-District Commands (Koramil). These Koramils oversaw security in East Timor's 464 villages, where the military posted a 'babinsa' (village guidance non-commissioned officer). Some villages had a village guidance team (Tim Pembina Desa, TPD). To this highly militarized structure were added two 'territorial' infantry battalions (Battalion 744 based in Dili and Battalion 745 based in Los Palos). Indonesia's military dominance persisted for the duration of the occupation.

International Support for Indonesia

Indonesia received considerable international support for its war against East Timor. With its vast natural resources, large population and strategic location along the main sea and air lanes between the Indian and Pacific Oceans, Indonesia was a country of great geopolitical significance. It was a founding member of the Non-Aligned Movement. It had the largest Muslim population in the world and was therefore very influential in the Organization of Islamic Conference. It was also a member of the Association of Southeast Asian Nations, which had a policy of non-interference in one another's internal affairs. Although the UN Security Council affirmed East Timor's right to self-determination in two resolutions in 1975 and 1976, it did not impose sanctions on Indonesia. The Australian government went out of its way to protect Indonesia from international criticism. Australian diplomats at the United Nations reported that their 'immediate diplomatic problem and task' was 'to do what we can to reduce the pressure on the Indonesians.' Australia's Deputy Representative to the United Nations, Duncan Campbell, expressed 'understanding and sympathy for Indonesia's position,' saying that Indonesia was 'inevitably touched and troubled by the tragedy in Timor.' Australia's Mission to the UN reported that 'Campbell's skilled and pertinacious negotiation in the Fourth Committee has kept the 'ASEAN plus' group together and a relatively mild resolution seems to be emerging which will … avoid condemnation of Indonesia [and] avoid recognition of the so-called Democratic Republic [of East Timor]' (NAA 1975b).

There were eight United Nations General Assembly resolutions concerning East Timor from 1975 to 1982. These were adopted with a declining majority of votes. The United States, Britain and France did not support any General Assembly resolutions from 1975 to 1982. The United States's position was subsequently explained by its Ambassador to the UN, Daniel Patrick Moynihan, in his memoirs:

> The United States wished things to turn out as they did, and worked to bring this about. The Department of State desired that the United Nations prove utterly ineffective in whatever measures it undertook. This task was given to me, and I carried it forward with no inconsiderable success.
>
> (Moynihan 1978: 279)

Only about one-third of the UN General Assembly, largely Third World states, kept the question of East Timor alive in the General Assembly from 1976 until 1982 when the matter was delegated to the UN Secretary-General. Only four Western states (Cyprus, Greece, Iceland and Portugal) supported East Timor in this period.

International Solidarity for East Timor

International solidarity was crucial to the Timorese quest for self-determination. There were other self-determination struggles in the Indonesian archipelago: Aceh, South Moluccas, West Papua. Only East Timor developed significant international solidarity, and only East Timor succeeded. A key activist throughout the period from 1975 to 1999 was Arnold Kohen, who was a 25-year-old volunteer journalist at an alternative radio program in Ithaca, New York in 1975. Kohen heard about the invasion through a range of sources, including taped press conferences by FRETILIN representative Jose Ramos-Horta, another 25-year-old, who had left East Timor with a few colleagues just days before the invasion to mobilize international support for the cause. Ramos-Horta and Kohen developed connections that endured for decades. Ithaca was the home of Cornell University, whose Modern Indonesia Project was the most influential Indonesian studies program in the world. Kohen was therefore able to contact one of the world's leading scholars of Indonesia, Professor Benedict Anderson, who agreed to be his advisor. This relationship also endured for decades, as did connections with other influential academics such as Professor Noam Chomsky of Massachusetts Institute of Technology. Chomsky's profile brought the East Timor question into universities around the world, informing many people about the atrocities and their misrepresentation by governments and the media (Kohen 2004).

One aspect of the campaign involved Congressional hearings into United States policy towards Indonesia and East Timor. Four hearings were held between March 1977 and February 1978—an extraordinary level of attention over an 11-month period for a small, peripheral territory like East Timor. The driving force behind these hearings was Dr John Salzberg, an aide to Representative Donald M. Fraser of Minnesota, who was chair of the Subcommittee on International Organizations of the Committee on International Relations. These hearings created a public record of Indonesian atrocities (US Congress 1977a, 1977b). They shattered the credibility of US officials during the famine of 1978–1979, when their previous testimony that the struggle had ended was contradicted by evidence of starvation and mass deaths in East Timor. For the same reason, the credibility of activists like Arnold Kohen was greatly enhanced. This credibility grew during the 1980s as a result of painstaking efforts to develop bipartisan initiatives in Congress, and outreach over many years to editorial writers and news staff of the New York Times, the Christian Science Monitor, the Washington Post and the Boston Globe, among others (Fernandes 2011: 67–70). Much of the pressure the US government exerted on Indonesia in 1999 had its origins in the structure of legitimacy grounded in these very important constituencies from the late 1970s through the 1990s. Activist credibility was further enhanced by the involvement of Catholic clergy with first-hand, ongoing knowledge of conditions in occupied East Timor. They were brought to the US Congress and other forums by Arnold Kohen. During the anti-communist atmosphere of the Cold War, they were an indispensable element in establishing East Timor's legitimacy with Congress, the news media and—not least—the US Catholic Church itself, which became a vital supporter over the next twenty years. Indeed, the US Catholic Church's links across the mainstream political spectrum helped bring about the change in US policy in 1999 that led to Indonesia's withdrawal.

Most of the international reaction to Indonesia's invasion took place in Australia. There was some public support for East Timor because Australian soldiers had fought there against Japan during World War II, with the help of many East Timorese. Another source of popular opposition to Indonesia's invasion arose as a result of the deaths of five journalists working for Australian TV stations. They were killed in the small border village of Balibo. Indonesia claimed the journalists had died in the heat of battle. Australia's ambassador to Indonesia reported that 'the truth of this incident may never be established.' He said the 'agreed consensus' was that the Australian Government should 'contain the damage to the long term Australian/Indonesian relationship' despite 'the very hostile media reaction to Indonesia's actions which is no doubt influenced by the Balibo affair' (NAA 1976b). But Australian activists kept up the pressure, and the secrecy surrounding their deaths—along with Indonesia's denials and Australia's prevarications—remained an important mobilising tool. Despite the best efforts of the Australian government, East Timor remained a running sore in its relationship with Indonesia.

The East Timorese Resistance

Indonesia's military superiority was never in doubt after the United States replenished its arsenal in 1977. But its operations did not completely eliminate FRETILIN's resistance. The most high-profile resistance leaders were killed but many cadres went to ground and lived to fight another day. They built a clandestine resistance to support a small band of guerrillas, who continued to harass Indonesian lines of communication. This armed resistance did not aim to evict Indonesian forces directly—it was too small for that. But its very existence implied a rejection of Indonesian rule and popular support for self-determination. There was no other way for guerrilla activity to persist over so many years in a small, occupied, place like East Timor, which lacked a land border with a friendly state, an external supplier of weapons, and a liberated area in which to recover between guerrilla operations. UN-sponsored negotiations between Portugal and Indonesia could not avoid the reality of the resistance, which served as a symbol to the outside world that a cohesive liberation movement existed inside the territory. 'To resist is to win' became a motto of the East Timorese resistance (Gusmao 2000).

Meanwhile, a new generation of East Timorese joined the resistance. They had grown up under the occupation, had no experience of Portuguese rule and little interest in continuing the 1975-era arguments between FRETILIN and its other Timorese rivals. Known as the *juventude* ('youth'), the new generation spoke Indonesian, were educated in the Indonesian school system, learnt the Suharto regime's version of Indonesian history, but were committed to an independent East Timor. The *juventude* came into their own in the second half of the 1980s when they received Indonesian government scholarships to study law, engineering or public administration at institutions of higher learning in East Java such as Malang, Yogyakarta, Semarang, Bali, Surabaya, Kediri, as well as in Jakarta and on the island of Sumatra. The Indonesian government established the East Timorese Students and Youth Association (IMPETTU) to monitor and ultimately assimilate them. Their membership in IMPETTU was compulsory (CAVR 2005: 269–271).

In order to counter the Indonesian government's aim of assimilating them, a group of nine East Timorese students studying in Bali formed the National Resistance of East Timorese Students (RENETIL) on 20 June 1988. RENETIL itself was an illegal organisation, but its members took over the leadership of IMPETTU and used it to run their own clandestine networks. Lawful IMPETTU sports competitions during the day were combined with illegal RENETIL evening meetings. This activity led to the formation of clandestine cells in Bandung, Semarang, Solo, Jakarta, Malang and Denpasar (Saky 2013). The students internationalized the struggle by conveying evidence of human rights violations in East Timor to international human rights organizations. They worked with clandestine activists inside East Timor to coordinate the visits of foreign diplomats and activists to the territory. They also participated in the Indonesian pro-democracy movement, forging links with Indonesian human rights activists (CAVR 2005: 466).

The *juventude* were keen to preserve a separate identity in order to prevent themselves from becoming Indonesianized. They had a sense of being a beleaguered Christian minority in a Muslim-majority Indonesia. They built links with Indonesian Christians from Flores, Solor and elsewhere in eastern Indonesia, to prepare themselves for an anticipated wave of fundamentalist Muslims from elsewhere in Indonesia. They were part of a movement called Hati Suci (Sacred Heart), in which they steeled themselves—physically and psychologically—against Indonesians who were trying to intrude on their territory. They held weekend camps in the mountains around Bogor and Bandung. They would walk for hours in bare feet to toughen themselves, navigate across country in the dark and do rigorous physical exercises (Fernandes 2011: 128-9). A Christian influence permeated their identity. Students who joined RENETIL swore an oath to East Timor's national liberation in a quasi-Christian ceremony; they would drink red wine to signify Christ's blood and would often pray for the Virgin Mary's intercession (personal communication with Aderito Soares, Head of Renetil's Political Analysis section, January 2009). At nationwide cultural events sponsored by the Indonesian government, they performed traditional Portuguese dances as a way of demonstrating their distinct identity. RENETIL's oath was written in Portuguese, and never translated into any other language.

The *juventude* participated in demonstrations inside East Timor and in Indonesia, and many were captured, interrogated, tortured, imprisoned or killed for their convictions. The most prominent event inside East Timor was the Santa Cruz massacre of 1991, when Indonesian forces opened fire on unarmed students who were marching from a church to a cemetery in the Santa Cruz neighbourhood of Dili. The massacre was captured on film by a foreign journalist whose visit had been facilitated by the resistance. It was broadcast worldwide to widespread international condemnation of Indonesia's conduct. Australia's foreign minister Gareth Evans tried to contain the damage by calling the massacre 'an aberration, not an act of state policy' (Aarons 1992). The Indonesian government appointed a National Commission of Inquiry to investigate. It assigned most of the blame to the students, saying that the shooting was a spontaneous reaction by Indonesian troops who were trying to defend themselves (Krieger 1997). Gareth Evans weighed in again, saying that there was 'no case to be supremely critical' (AFP 1992). The Indonesian government had responded to the killings 'in a reasonable and credible way' and therefore 'essentially punitive responses from the international community are not appropriate' (AFP 1992). Despite Evans' rhetoric, the Santa Cruz massacre showed that 16 years after the invasion, East Timor had not accepted Indonesian rule.

The most prominent event involving the *juventude* outside East Timor was their occupation of the US Embassy in Jakarta during the November 1994 APEC summit. Twenty-nine students climbed over the 2.6 metre railings and jumped into the embassy compound before security guards positioned outside could stop them. They unfurled banners and chanted 'Free East Timor.' Dozens of riot police arrived soon afterwards and surrounded the embassy. A stand-off began in full view of the international media, and the cause of East Timor was dramatically reasserted onto the international agenda. The students stayed in the Embassy for twelve consecutive days, rejecting numerous offers of safe passage and asylum. Only after they had ensured that the international media's focus during APEC was on East Timor rather than on Indonesian President Suharto's 'accomplishments' did they accept an offer of asylum to Portugal.

Indonesia's military presence remained all-pervasive within the territory, however. Its troops captured the commander of the resistance, Xanana Gusmao, in November 1992, and the man who succeeded him, in April 1993. They captured the leader of the clandestine resistance in 1995. Indonesia's intelligence agents penetrated the resistance and sowed doubts among its members. The level of suspicion and paranoia increased because of the large number of collaborators and double agents. According to a foreign activist, 'You would be in very secret meetings with only four members of the clandestine resistance and not know which of the four was betraying you' (communication with Toshihide Nakayama, January 2008). A resistance activist later recalled that when he was arrested he was taken to the house of the military commander, who spoke to him at length about the extent of Indonesian infiltration. He said, 'I was surprised … because he talked about the movement's activities in Java as if he were East Timorese. He knew everything. I mean, all about the movement's activities in Java and its ties with Dili, the Resistance groups, the student organisations and other organisations in Java' (CAVR 2005: 1620). The oppressive military presence had an additional consequence of preventing meaningful, broad-based discussion of contemporary politics or political arrangements in a future, independent East Timor, since no such gatherings of East Timorese were permitted.

A Continuum of Resistance

Despite the intense pressure, a robust continuum of resistance developed between the armed fighters in the mountains, the clandestine resistance in the towns, the students on scholarships in Indonesia, the East Timorese diaspora abroad, and international activists. This continuum of resistance was the most decisive factor in East Timor's war of independence. It ensured that the war was fought along multiple dimensions: in the mountains, towns and villages of East Timor, in the cities of Indonesia, at the United Nations, in a number of countries around the world, and in the international media. The weapons used by the combatants were not just bullets but newspaper articles, public talks, films, texts and protests in order to weaken international support for Indonesia's occupation. The resistance enjoyed dynamism and flexibility as a result of the war's multiple dimensions. Although outnumbered in many ways, a setback in one theatre did not mean the end of the struggle.

Events in one theatre of war were quite consciously designed to influence events in another theatre. Protests in East Timor were arranged to coincide with Portugal-Indonesia negotiations occurring on the other side of the world. The clandestine resistance or visiting foreign activists filmed these protests and ensured they gained political traction when broadcast by overseas television stations. International publicity campaigns featured speeches by diaspora East Timorese and their foreign supporters (Webster 2020). These campaigns were aligned with the Indonesian pro-democracy movement; they could not be accused of being 'anti-Indonesian' because they welcomed Indonesian participation, called for the overthrow of the New Order regime and the establishment of an independent East Timor alongside a democratic Indonesia. Pro-democracy Indonesian activists made self-determination for East Timor an integral part of their campaign as well (Wilson 2010).

The resistance also benefited from a conscious decision to avoid violence against non-combatants. It did not target Indonesian civilians in East Timor, nor Indonesian government officials in the wider world. Indonesian diplomats in Australia, for example, were unable to point to violent actions against themselves or their premises. Activists would attend talks given by the Indonesian ambassador, ask questions, hand out leaflets outside the event, and hold their own educational events. But they never threatened or used violence against Indonesian officials—or against Australian officials, for that matter. There was a clear demarcation between the right of the armed resistance to defend themselves against the Indonesian military inside East Timor and the nonviolence of the rest of the campaign. Such restraint made it much easier for Australian politicians, religious figures, human rights campaigners and other civil society groups to openly champion the cause of East Timor. Public support for East Timor's right to self-determination remained strong.

The legitimacy of East Timor's cause received a major boost at the end of 1996 when the Nobel Peace Prize was awarded to Jose Ramos-Horta, by then a resident of Australia, and Father Carlos Belo, the apostolic administrator of the Catholic diocese of Dili. This was a huge victory for the independence campaign. Activists recognized that the Prize 'created something uncontrollable because after that the issue couldn't simply be buried. East Timor would never go away now without a political settlement' (conversation with Arnold Kohen, March 2008, November 2010). Ramos-Horta's long-term advisor later observed that the Nobel Peace Prize 'was probably *the* turning point.' Before the Nobel Prize, 'nobody really wanted to see us ... It was undeniable that East Timor was a human rights issue. But nobody was really terribly interested or committed to it. They had to receive us but it was clear that they wished we went away'. After the Prize, 'all of a sudden all sorts of invitations appeared ... Everybody wanted to have a Nobel Laureate decorating their tables, their dinners, their meetings, their conferences and their events because it gave them prestige' (interview with Juan Federer, January 2008). Indonesia's foreign minister Ali Alatas later said, 'It was as if we had been besieged. The 1996 Nobel Peace Prize showed how seriously we had been besieged' (Alatas 2000).

By 1998, therefore, Indonesia retained its military superiority but international support for its occupation of East Timor was less secure. Australia was now the centre of a strong pro-East Timor solidarity movement. The Labor opposition now called for East Timorese self-determination, ending a quarter of a century of bipartisan political support for Indonesia's occupation. The East Timorese resistance was surging. It was in a good position to capitalize on any vulnerability in Indonesia's armour. This vulnerability appeared in the middle of 1997 in the form of the Asian Financial Crisis, which led quickly to the implosion of the Indonesian economy. Massive public protests forced President Suharto to resign on 21 May 1998. His successor was Dr B J Habibie, an aeronautical engineer and chairman of the Association of Indonesian Muslim Intellectuals (ICMI). Members of ICMI had varying preferences about the future of Indonesia. Some had a populist vision of the future. Others believed in empowering small businesses. Still others, like President Habibie, believed in a high-technology future. On the future of East Timor, however, most senior ICMI personnel saw no compelling reason to retain it within Indonesia.

The Fracturing of Elite Indonesian Views Regarding East Timor

ICMI's Secretary-General was Adi Sasono, an engineer by training and Habibie's close ideological ally. He had visited East Timor several times and believed the Indonesian military's actions had created 'a totally unacceptable situation.' Expressing his view privately in 1993, he said there had been harm to Indonesia, too: 'The cost of international criticism, which is justified, is too high for Indonesia to afford. It shames Indonesia and has a negative effect on the international support needed for development.' This was, of course, due to the international solidarity network for East Timor, which constantly highlighted Indonesia's ongoing occupation and human rights violations. Sasono 'stressed' that ICMI understood 'the strength and validity of international opposition to Indonesia's East Timor policies, and the urgent necessity to remove this seriously high economic, political and social cost, which is unnecessary and only the result of misplaced military policies' Fernandes 2011: 117-8).

President Habibie had previously encountered the ferocity of international activist campaigns on East Timor. During an official visit to Australia while still Minister for Research and Technology, he had been subjected to repeated questioning about East Timor—including by journalists at the National Press Club in Canberra—almost to the exclusion of any other issue. The East Timor question dogged him wherever he went. His closest adviser on foreign relations, Dewi Fortuna Anwar, understood that 'East Timor had become a real albatross.' Despite 'Indonesia's diverse assets, and social, political and economic capital, our diplomats had to spend all the time being defensive.' Habibie had to deal with the effects of the Asian Financial Crisis as soon as he became president. Real GDP was 16.5% below the same period in 1997. The exchange rate was more than four times lower than the previous year. Imported goods had become prohibitively expensive. Inflation soared and the purchasing power of the rupiah was plummeting (McGillivray 1999: 3–26). Wage-earners had lost more than a third of their real incomes. Domestic unrest was threatening to get out of control. To compound all this, oil prices—a key source of government revenues—were stagnating at $10–12 per barrel. And yet, according to Anwar, 'the moment he became President, all the questions from international visitors were about East Timor.' This led to the feeling that 'the mothership is in danger of sinking. We really need to shed the excess baggage' (conversation with Dewi Fortuna Anwar, 10 June 2015).

The technocrats in Habibie's cabinet were of a similar view. Ginandjar Kartasasmita, the Coordinating Minister for the Economy, believed that East Timor deserved the right to self-determination. ICMI's Adi Sasono had entered the cabinet as Minister for Cooperatives. They had played no part in the decision to invade East Timor. They resented having to bear the burden of a policy for which they were not responsible, with which they did not agree, and from which they derived no benefit. Buffeted from all sides, Habibie announced on 24 July 1998 that 1,000 combat troops would be withdrawn from East Timor as an act of goodwill (Reuters 1998). The East Timorese resistance exposed the truth about these announcements. They had infiltrated the Indonesian military's headquarters and obtained its entire list of personnel, units, and pay records. They provided it to an Australian activist named Andrew McNaughtan, who smuggled it out of the territory and—after analysis—showed that Indonesia's claims of withdrawal and demilitarization in East Timor were lies (Greenlees 1998). The number of combat troops had increased to 40% of the total figure (AAP 1998). All the militia groups in East Timor were under the military's direct command. Military personnel occupied all the key civilian positions in East Timor (KODAM IX 1999: 279).

The negative publicity surrounding these revelations ratcheted up the pressure on Habibie. Determined to cut East Timor loose, he discussed the possibility of holding a vote on independence (Greenless and Garran 2002: 87–101). He consulted two cabinet members with a military background: Major General Sintong Panjaitan and Lieutenant General Feisal Tanjung. Both had been members of the military team that manipulated the fraudulent Act of Free Choice in West Papua in 1969. They were familiar with the process of engineering desired ballot outcomes. Tanjung later said that the Indonesian military believed about 75% of East Timorese would vote to remain in Indonesia (Hisyam 1999: 219–232, 724, 738). Foreign Minister Ali Alatas later said that most members of the cabinet 'were then very convinced we would win the referendum. Everything was painted with optimism' (Alatas 2000). Encountering no opposition from Cabinet, the Indonesian government announced on 27 January 1999 that East Timorese would be granted a referendum on independence. Thus ended Indonesia's elite consensus that its 1976 annexation of East Timor was irreversible.

State-Sponsored Terrorism and East Timor's Popular Consultation

The referendum on autonomy versus independence was called a 'popular consultation.' It was held under the auspices of the United Nations, which began deploying to East Timor at the end of May 1999 as the United Nations Mission in East Timor (UNAMET). Electoral staff, unarmed civilian police and unarmed military liaison officers from around the world were deployed to all 13 districts alongside local East Timorese staff. The Indonesian military tried to rig the outcome in favour of special autonomy within Indonesia. The military aspect of this strategy was to intensify operations under the guise of proxy forces, known as the 'militia'. The strategy also had a political component: minimizing international involvement whilst it carried out these militia operations. In this latter aspect Indonesia received valuable assistance from the Australian government, which repeated the Indonesian military's claim that it was merely keeping the peace between rival East Timorese factions: its militia proxies on the one hand and pro-independence campaigners on the other (Downer 1999).

Over the course of 1999, the Indonesian military directed its militia proxies to attack and intimidate the people of East Timor. The aim of this campaign of state-sponsored terrorism was to leave the East Timorese in no doubt that a vote for independence would result in even more terror. A more ambitious objective was to have the popular consultation cancelled altogether, on the grounds that the violence was not conducive to the electoral process. The people of East Timor heard the message clearly—but insisted that the ballot be conducted anyway. The resistance had formed the view that there would not be another chance to overthrow Indonesian rule. It was now or never. After Indonesia recovered from the financial crisis and the world's attention had moved on from a small territory like East Timor, the Indonesian authorities would simply reassert control and declare that their rule over the territory was irreversible.

The international community's presence in East Timor was a major barrier to Indonesia's strategy. A number of foreign governments maintained a close watch on the situation. UNAMET accredited some 600 journalists over the course of 1999. It accredited around 1,700 East Timorese and Indonesian non-governmental observers, and an additional 2,300 international observers to monitor the referendum. Around 500 international observers also arrived with government delegations representing Australia, New Zealand, Brazil, Canada, Chile, Ireland and Spain, as well as the European Union. As East Timor's Truth Commission later noted, the 'international presence was unprecedented in the history of [East] Timor. In the twenty-five-year period of Indonesian presence it had been unthinkable that the territory could have been so open to the international community' (CAVR 2005: 289). Their presence 'across the territory provided a level of monitoring that may have contributed to the reduction in large-scale attacks' by the Indonesian forces (CAVR 2005: 289).

The international presence supported the resolute defiance of the East Timorese people, defeating Indonesia's goal of cancelling the ballot. The vote took place on 30 August 1999. The results were announced on Saturday 4 September 1999. 78.5% of registered voters opted for independence from Indonesia. The result took the Indonesian authorities by surprise. They had expected to win comfortably, not lose by such a large margin. Two decades of authoritarian control over East Timor had blinded them to the truth of East Timorese public opinion. They moved swiftly to try to reverse the result of the ballot. They used more violence and intimidation to remove the thousands of foreign observers from the territory. They forcibly deported up to 250,000 East Timorese—one-third of the population—across the border to West Timor (KPP-HAM 1999: 15–59). They deployed assassination squads to hunt down and kill the leadership of the independence movement (McDonald 2002).

The Indonesian aim now was to discredit the result of the ballot as rigged, by suggesting that the East Timorese now in West Timor had voted with their feet in accordance with their true wishes. Having gotten away with terror for so long, they expected things would be dealt with in the usual fashion—expressions of regret by the United States and Australia but no practical punitive action. Here, too, they failed. One reason was the disciplined restraint by the armed resistance, who were in any case heavily outnumbered. They refused to hit back directly at those carrying out the terror campaign, depriving Indonesia of a pretext to send even more troops to saturate the territory. A second reason was that the United Nations had guaranteed that the popular consultation would take place and that its outcome would be honoured. Reversing the ballot through naked terror would harm the credibility of the United Nations for future electoral processes in other parts of the world. A third reason was that the Australian government, a long-time supporter of the occupation, was confronted with a tidal wave of outrage from its own population (Fernandes 2004: 86–114).

The major reason Indonesia failed, however, was that the sole superpower stepped in. The Clinton administration had come under heavy pressure to act as the international outcry over the carnage in East Timor and its own paltry response as it reached a crescendo. Influential domestic constituencies demanded action. Liberal human rights campaigners and politicians were joined by conservative religious and political figures. The United States Conference of Catholic Bishops made forceful representations to the Clinton Administration between 9 and 15 September 1999 (Kohen 2004). Conservative Catholic Republicans led by William P Clark, former President Reagan's national security advisor, made it clear that there was support across the political spectrum for a decisive response. Numerous commentators in the media amplified this sense of outrage. The Australian government—facing its own domestic uproar as a result of its decades-long support for Indonesia—appealed to the United States, its long-term military ally. The Clinton administration finally realised that US policy had to be reversed.

How It Ended

In an atmosphere of mounting international outrage, the United States finally stepped in and informed the Indonesian generals that the time had come to withdraw. Admiral Dennis Blair, Commander-In-Chief of US forces in the Pacific, met Indonesian Defence Minister and Chief of the Armed Forces, General Wiranto, in Jakarta on 8 September 1999. The records of the meeting indicate that Blair urged General Wiranto to 'pull back from the brink of disaster' and to provide 'immediate evidence' of having done so:

> Despite repeated assurances that the Indonesian armed forces could fulfil its obligations to maintain security in East Timor, despite sending substantial numbers of new troops to the territory and taking the extraordinary step of imposing martial law, East Timor has descended into anarchy … Further deterioration of the situation will not only cause unnecessary loss of life, it will do potentially irrevocable damage to Indonesia's relationship with the rest of the world, including the US. As you know, a coalition of concerned nations is willing to send a multinational force to East Timor; such a force would aim to stabilize the

> situation until the MPR [the legislative branch in Indonesia's political system] meets to endorse the outcome of the election, then new arrangements would be made with the UN. The whole world is watching as this tragedy unfolds, and international condemnation of Indonesia has grown to a fever pitch. The window of opportunity in which Indonesia can salvage its relations with the world is rapidly shutting.
>
> (Barker 2019)

Admiral Blair briefed President Clinton on the East Timor situation after this meeting. Clinton then announced that Indonesia 'must invite—it must invite—the international community to assist in restoring security. It must allow international relief agencies to help people on the ground. It must move forward with the transition to independence. Having allowed the vote and gotten such a clear, unambiguous answer, we cannot have a reversal of course here' (Clinton 1999a). A few hours later, whilst aboard Air Force One on the final leg of a flight to Auckland for the Asia Pacific Economic Cooperation (APEC) summit, Clinton issued his strongest statement yet, accusing the TNI of direct involvement: 'It is clear the Indonesian military is aiding and abetting the militia violence ... This is simply unacceptable' (Clinton 1999b). International pressure mounted on Indonesia during the APEC summit, with Canada and New Zealand removing all sources of comfort for Indonesian policymakers.

The Indonesian military's resistance ended within hours. On 12 September 1999, Habibie and Wiranto emerged from a special Cabinet meeting and agreed to accept an international peacekeeping force into East Timor. Indonesia made a last-ditch attempt to exclude Australia by seeking to determine the composition of the force, but this was quickly resolved by the US's public insistence that Indonesia should 'not be able to say who is in or not in the force and what the structure of the force would be' (McDonald 1999: 1). All senior Indonesian generals met at the Golf Club at the Armed Forces headquarters in Jakarta immediately afterwards. They accepted that Indonesia would not oppose the lodgement of the peacekeeping force. The International Force—East Timor (INTERFET) under Australian leadership entered the territory on 20 September 1999.

To ensure there would be no lower-level initiatives to attack INTERFET, US Secretary of Defense, William Cohen, met General Wiranto shortly afterwards. According to a State Department cable, Wiranto began by claiming the situation in East Timor had been 'greatly exaggerated by the media' and that things were now under control. He continued to deny that his military had backed the militias. Cohen responded by saying he would 'have to be quite direct.' Indonesian dissatisfaction at the results of the polls 'could never justify the rampage which had followed.' It was widely believed TNI supported the militia (supported by first-hand visual evidence), and Cohen was also 'concerned about indications that the militias in West Timor were considering ways in which they might be able to threaten INTERFET.

Such actions would be tragic if they were allowed. It was imperative that Wiranto show leadership in preventing such actions.' Cohen warned Wiranto that a 'positive bilateral relationship … would not be possible unless Indonesia made progress' (Barker 2019).

Cohen's reading of the riot act marked the loss of Indonesia's will to occupy East Timor. The Indonesian high command understood that the game was up. One of the most unlikely events in regional history had occurred. During World War II, a senior advisor to US President Roosevelt observed that the territory might eventually achieve self-government, but 'it would certainly take a thousand years' (Louis 1987: 237). East Timor's aspirations for independence were widely regarded as impossible; it did not have a land border with a friendly state or an external supplier of weapons, and its fighters did not have a liberated area in which to recover between guerrilla operations. Given these conditions, its successful resistance is probably unique in the history of guerrilla warfare and independence struggles.

Notes

1 **Dedication:** I dedicate this paper to the late Dr Andrew McNaughton
2 Today known as Timor-Leste, it was called East Timor by most people during the period covered in this chapter.

References

AAP, 1998. 'Indonesia admits 18,000 troops may be in East Timor.' *Australian Associated Press*, A newspaper. 30 October.

Aarons, M. and Domm, R., 1992. *East Timor: A Western-made Tragedy*. Left Book Club, Sydney.

AFP, 1992. 'Australia critical of U.S. decision to penalise Indonesia.' *Agence France-Presse*, A newspaper. 26 June.

Alatas, A., 2000. 'Santa Cruz incident a turning point in our diplomacy.' *Tempo* 18-24 September. vol. 1, no. 1. (English language edition).

Barker, A., 2019. 'Declassified intelligence documents shed light on 1999 Timor-Leste independence,' *ABC News*, A news website. 29 August.

CAVR, 2005. *Final Report of the Commission for Reception, Truth and Reconciliation*. Timor-Leste Comissao de Acolhimento, Verdade e Reconciliacao (CAVR), Dili.

Clinton, W.J., 1999a. *Statement on East Timor—Washington, DC*. Voice of America, 9 September 1999.

Clinton, W.J., 1999b. *East Timor Statement—Hawaii*. Voice of America, 9 September 1999.

Downer, A., 1999. *Face To Face program*. Channel Seven, 21 February.

Fernandes, C., 2004. *Reluctant Saviour: Australia, Indonesia and the independence of East Timor*. Scribe Publications, Carlton.

Fernandes, C., 2011. *The Independence of East Timor: Multi-dimensional perspectives*. Sussex Academic Press, Eastbourne.

Greenlees, D., 1998. 'Leak shows no East Timor troop cuts', *The Australian*, A newspaper. 30 October.

Greenlees, D., & Garran, R. (2002). Deliverance: The inside story of East Timoras fight for freedom, Allen and Unwin, Sydney.

Gusmao, X., 2000. *Resistir E Vencer! (To Resist is to Win!): The Autobiography of Xanana Gusmao*, Niner, Sara (ed). Aurora Books, Richmond.

Hill, H., 1980. 'The NGOs and East Timor', East Timor Today, ACFOA Development Dossier No 1 July 1980, Australian Council for Overseas Aid.

Hisyam, U., 1999. *Feisal Tanjung: Terbaik Untuk Rakyat, Terbaik Bagi ABRI*. Dharmapena, Jakarta.

Hoadley, J., 1977. 'Indonesia's Annexation of East Timor: Political, Administrative and Developmental Initiatives', in Kim Kahn Huynh (ed.), *Southeast Asian Affairs 1977*. ISEAS Publishing, Singapore.

KODAM IX, 1999. *Udayana, Komando Resor Militer 164, Rekapitulasi Kekuatan Personil Organik dan Penugasan, July 1998; Rekapitulasi Kekuatan Personil Organik dan Penugasan Posisi, November 1997*. Komando Daerah Militer IX, Dili.

Kohen, A., 2004. *Testimony given to the CAVR National Public Hearing on Self-determination and the International Community*, 15–17 March. CAVR, Dili, 2013.

KPP-HAM, 1999. *Full Report of the Investigative Commission into Human Rights Violations in East Timor*. Komisi Penyelidik Pelanggaran Hak Asasi Manusia di Timor Timur, 31 October, Jakarta.

Krieger, H. (ed.), 1997. 'National Commission of Inquiry into 12 November 1991 Incident in Dili, 26 December 1991', *East Timor and the International Community: Basic Documents*. Cambridge University Press, Cambridge.

Louis, W.R., 1987. *Imperialism at Bay: The United Stated and the Decolonization of the British Empire, 1941-1945*. Oxford University Press, Oxford.

McDonald, H., 1999. 'Conditions on foreign force could limit Australia's role', *Sydney Morning Herald*, A newspaper. 13 September.

McDonald, H., 2002. 'Silence over a crime against humanity', *The Age*, A newspaper. 14 March.

McGillivray, M. and Morrissey, O., 1999. 'Economic and Financial Meltdown in Indonesia: Prospects for Sustained and Equitable Economic and Social Recovery', in Budiman, Arif et al, (eds), *Reformasi: Crisis and Change in Indonesia*. Monash Asia Institute, Melbourne.

Moynihan, D.P., 1978. *A Dangerous Place*. Berkley Books, New York.

Murtopo, A., 2003. '25 Years of Accelerated Modernisation', in Bourchier, David and Hadiz, Vedi (eds), *Indonesian Politics and Society: A Reader*. Routledge, London.

NAA, 1974. 'Loose papers of the South-East Asia Branch, Correspondence files of the Departments of External Affairs and Foreign Affairs. National Archives of Australia: A11443, [1]'.

NAA, 1974-5. 'Submissions to Ministers and briefs on Portuguese Timor. National Archives of Australia: A1838, 3038/10/2, ii'.

NAA, 1975a. 'Joint Intelligence Organisation, A descriptive survey of Portuguese Timor, February 1975. National Archives of Australia: A13685, 3/75'.

NAA, 1975b. UNGA 30: 4th Committee: May-Dec. 1975. National Archives of Australia: Al838, 906/30/14/3, ii.

NAA, 1976a. Joint Intelligence Organisation, Military Study—Indonesia, Part 1. National Archives of Australia: A13685, 12/1976.

NAA, 1976b. Correspondence files, Djakarta/Jakarta; East Timor. National Archives of Australia: A10463, 801/13/11/1, xvii.

NAA, 1977. Joint Intelligence Organisation, *Indonesia Military Study,* March 1977. National Archives of Australia: 13685.

NAA, 1978. The Indonesian integration of East Timor. National Archives of Australia: A13685, 1/1978.

Nevins, J., 2005. *A not-so-distant horror: Mass Violence in East Timor*. Cornell University Press, Ithaca.

Peake, G., 2012. 'An Increased Spotlight: Australia in Timor-Leste', *Security Challenges*, vol. 8.4.

Power, S., 2002. *A Problem From Hell: America and The Age Of Genocide*. Basic Books, New York.

Reuters, 1998. 'Some Indonesian troops to leave East Timor,' 24 July.

Saky, C. da S., 2013. *RENETIL iha luta libertasaun Timor-Lorosa'e*. Resistência Nacional dos Estudantes de Timor-Leste (RENETIL), Dili.

Staveteig, S., 2007. *How Many Persons in East Timor Went 'Missing' During the Indonesian Occupation?: Results from Indirect Estimates*. International Institute for Applied Systems Analysis, Austria.

US Congress, 1977a. *Economic and Military Assistance in Asia and the Pacific*, US Congress House of Representatives, 10, 17 and 22 March.

US Congress, 1977b. *Human Rights in East Timor and the Question of the Use of US Equipment by the Indonesian Armed Forces*, US Congress House of Representatives, 23 March, 23 June, 28 June and 19 July 1977.

US Congress, 1977c. *Human Rights in East Timor*, US Congress House of Representatives, 19 July.

US Congress, 1978. *US Policy on Human Rights and Military Assistance*, US Congress House of Representatives, 15 February.

Wanandi, J., 2012. *Shades of Grey: A Political Memoir of Modern Indonesia 1965-1998*. Equinox, Jakarta.

Weatherbee, D., 1966. 'Portuguese Timor: An Indonesian Dilemma,' *Asian Survey*, vol. 6.2, pp. 683–695.

Webster, D., 2020. *Challenge the Strong Wind: Canada and East Timor 1975-1999*. UBC Press, Vancouver.

Wilson, I., 2010. *A luta continua!: Politik radikal di Indonesia dan Pergerakan Pembebasan Timor Leste*. Tanah Lapang, Jakarta.

6 Ending The Sierra Leone War

Richard Iron

The 1991 to 2002 war in Sierra Leone conjures a melange of horrific images: amputations and mass rapes; abduction of children and child soldiers; blood diamonds; and the downfall of Charles Taylor, sentenced in 2012 to fifty years' imprisonment for being 'responsible for aiding and abetting as well as planning some of the most heinous crimes recorded in human history' (SCSL 2012).

The overall impression of the war is its nihilism: its lack of purpose and the random perpetration of extreme violence. But to regard it purely as nihilistic is to forego the opportunity to understand what happened, why it happened and why it developed in the form it did. It is only by achieving this understanding that we can hope to learn something from how the war ended.

One of the reasons for a lack of understanding of the war in Sierra Leone is its complexity. There were three guerrilla organisations: the Revolutionary United Front (RUF), the Armed Forces Revolutionary Command (AFRC) and the Civil Defence Force (CDF)—the latter fighting on the side of the legitimate Government of Sierra Leone. There was also coup and counter-coup; four peace treaties; a significant involvement of the West African regional organisation ECOWAS and its military arm ECOMOG, led by Nigeria; the UN's largest ever military operation; involvement by private military companies; and, finally, a limited but decisive intervention by the previous colonial power, the United Kingdom. Additionally, Sierra Leone felt the baleful effect of Muammar Gaddafi's Libya, where the leaders of revolutions in Sierra Leone (Foday Sankoh), Liberia (Charles Taylor), and Guinea Bissau (Blaise Campaoré) were all trained and supported with arms.

Why Did The RUF Fail?

Although there were many actors in Sierra Leone, the principal destabiliser was the RUF. For the war to end, the RUF had to stop, or be stopped from, fighting. Ultimately, of course, that occurred and the RUF was defeated. To understand how and why the war ended, it is important first to understand the RUF which had a number of particular characteristics as a guerrilla

DOI: 10.4324/9781003317487-9

organisation. Some are unique to the RUF; others are common to many African groups. There are five that are particularly important to understanding how the war evolved and, eventually, ended (Iron 2005).

The RUF was Neither A Popular Nor A Populist Movement

Although the RUF's leader, Foday Sankoh, attempted to woo public support for his invasion of Sierra Leone in 1991—indeed he attracted a significant number of volunteers in the first year of the war—the movement's depredations against the population soon lost it any measure of popularity among the urban and rural poor. Both are constituencies from which the RUF should have freely expected to draw supplies, recruits and intelligence. Sankoh's attempts to impose discipline on the organisation were doomed to fail: he had been loaned a powerful contingent of Liberian fighters from Charles Taylor; these were beyond his control and the population did not distinguish between Liberian or Sierra Leonean members of the RUF when committing abuses.

On the other side, democratic elections in 1996 that elected Ahmed Tejan Kabbah as president, together with government sponsorship of trusted local defence forces, went a long way to overcome the rural population's distrust of the Sierra Leone Army, which had also conducted depredations against the population in the early years of the war. This was coupled with significant international aid, channelled through the government, to improve the lot of both urban and rural populations. The national government thus largely retained the support of its population throughout the war, except for when a military junta temporarily usurped power in 1997–1998.

As a result, the RUF had to rely on fear to impose its will on the people it controlled or wanted to control, inflicting violence on the population, rather than having a widely accepted legitimacy to rule. The unpopularity of the RUF was quantified in the 2002 presidential elections, after the final ceasefire had been implemented and the country was at peace, when the RUF candidate attracted only 1.7% of the popular vote and President Kabbah was re-elected with 70.1%.

In any insurgency, the population is largely divided in its loyalty between the government and the insurgent. Many are somewhere in between, with perhaps a leaning towards one of the other. Success in insurgency or counter-insurgency is usually determined by how well each side manages to influence sections of the population towards, or away from, the insurgent movement or government.

No such struggle for the hearts and minds of the population occurred in Sierra Leone. The RUF largely saw civilians as a resource to be exploited by the abduction and recruitment of child soldiers, provision of forced labour and easy availability of sex slaves, rather than a prize to be won over to give the organisation a popular legitimacy. Thus, few of the population were adversaries to the government in their battle against the RUF. Although some resisted the Sierra Leone Army, as suspicious of army depredations as of the RUF's, it did not make them support the RUF: instead, most put their trust in the government-supported local militias that evolved into the Civil Defence Force.

The lack of the RUF's legitimacy meant that the organisation depended solely on military power to rule, rather than having any kind of moral authority. Defeat of its military wing meant defeat of the organisation as a whole. There were no popular demonstrations for the RUF or disruption of labour in Freetown; there was no student agitation for the RUF against the government; and there was no RUF-inspired rural terrorism in areas recaptured by government forces from the RUF.

For the Sierra Leone War to end, the RUF had to be militarily defeated.

The RUF had no Effective Ideology to See it Through Difficult Times

Most insurgent, guerrilla or terrorist groups have some form of ideology which forms the basis of a political programme. Such ideologies provide motivation to the group's members who remain convinced of the rightness of their cause even when things are not going well, such as when suffering military reverses. Ideology, and the political programme that it underpins, can also appeal to wider sections of the population, not directly involved in the insurgency, to build passive support for the insurgency and against the government.

The RUF's leadership was, seemingly, proud of its ideology. It produced, in about 1995, a document entitled *Footpaths to Democracy: Toward a New Sierra Leone* that purported to be the ideological basis of the movement (RUF 1995). But, as Professor Christopher Clapham pointed out, the document is oddly devoid of ideology: there is little or no reference to ideological antecedents with no mention at all of Marxist ideas of class conflict or, even more strangely, no quotation of Mao Zedong where one might have expected phrases such 'people's war' or 'protracted struggle' to appear. Indeed, the style and standard of writing in *Footpaths* indicate that it was written primarily for an international, rather than internal, audience and is likely to have been authored by people who were not residents in Sierra Leone at the time. The ideological programme, such as it existed, was limited to complaints about the venality of the Freetown government and the need to redistribute Sierra Leone's mineral wealth (Clapham 2004, personal communication, 3 July).

According to Gibril Massaquoi, at one stage Sankoh's spokesman, the RUF's fighters had no interest in *Footpaths* or the ideology it claimed to outline. He was tasked with visiting all RUF field units to explain the ideology but claimed it was a dispiriting experience with the commanders, in particular, being only interested in the conduct of the military campaign (G Massaquoi 2003, interview in Freetown, 3 November).

Without any form of underpinning ideology, the RUF had to find the motivation to keep fighting from within itself. Such self-motivation is common in all military organisations and is grounded in good leadership and small unit cohesion. The principal consequence of this was that a RUF fighter's loyalty was mainly to his or her own small fighting group rather than to the organisation as a whole, especially after Sankoh's arrest in Nigeria in early 1997. This made the RUF prone to in-fighting and factionalisation which resulted in near-civil war within the organisation for much of 1999.

One other important aspect of the RUF's self-motivation came from the sense of empowerment that RUF membership brought. After the war, most child soldiers when interviewed claimed that having an AK-47 gave them power: it made them feel important (for example P Kamara 2004, interview in Makeni, 2 February). Similarly, commanders in the RUF had far greater power and responsibility than most could ever hope for in peacetime (for example G Johnson alias 'Junior Lion', 2003, interview in Freetown, 10 October). Being an arbiter of life and death, on a whim, and free availability of sex from almost any woman they desired, made many young men and boys in the RUF feel important and empowered. It was this empowerment that was in large part responsible for the RUF's reluctance to give up its weapons and control of its people as the war drew to a close: a return to civilian life would have resulted in a significant loss of personal power and status.

Thus one of the most important terms of the Abuja Agreement, which finally ended the war, was the inclusion of RUF fighters in the DDR process and their subsequent absorption into Sierra Leone's new armed forces. Individual RUF fighters wanted to retain status and authority as a soldier. The British designed and managed Military Reintegration Plan, following the UN-sponsored DDR process, was important in giving the RUF the confidence that they would be treated fairly. In the end, only about 10% of the RUF were admitted into the new armed forces (many were too young), with very few officers, but the mechanism was an important part of bringing the RUF to peace (Dent 2002).

The RUF was Highly Dependent on External Support

Given that the RUF could not rely upon the support of the population, for much of the war it was largely dependent on external sources for munitions, cash and even food.

Throughout the war, the RUF was highly reliant on Charles Taylor and his Liberian guerrilla group the NPFL and, later, his Liberian government. This reliance started with the loan of NPFL fighters, continued with Liberia acting as a conduit for Libyan-supplied weapons, and developed as the RUF became increasingly reliant on Liberian-supplied ammunition for its major operations. Charles Taylor's principal interest in Sierra Leone was its diamonds and the RUF invested considerable effort in diamond mining to pay for Liberian support.

This had a number of major implications for the RUF and the war. First, and most important, it made the RUF complicit in Charles Taylor's criminal enterprise. Any legitimacy it may have gained internationally by fighting what might be seen as a corrupt and overbearing Freetown government was completely undermined by its involvement in the illicit blood diamond trade.

There were further operational implications for the RUF. The first was that they needed to establish effective occupation of at least some of Sierra Leone's diamond fields: this meant that, as a guerrilla organisation, they could not hide in the jungle from security forces but had to operate more like a conventional force and occupy territory. It also meant that they were reliant on a tenuous logistic umbilical cord through the Sierra Leone jungle, using enslaved labour to carry large amounts of ammunition from the Liberian border: as the RUF advanced closer to Freetown it found it more difficult to sustain sufficient ammunition flow and became decreasingly effective.

The RUF was Reliant on A Small Number of Experienced and Charismatic Field Commanders

By early 1999 the RUF was as tactically effective as it had ever been. On 16 December 1998 it mounted a complex and well-executed operation to destroy a Nigerian all-arms regular brigade in the diamond mining centre of Koidu. All its commanders and many of its fighters had been continuously engaged in combat for over seven years. The movement had mutated through conventional war and jungle-based insurgency: it had become an adaptable organisation with many of the systems and techniques more usually associated with a regular army. For example, it had its own provost staff, medical services, chaplains, radio communications specialists and logistic systems.

Nevertheless, the RUF tactical force was still largely made up of children who, when in combat, were often under the influence of narcotics. Their discipline, both in the jungle and in combat, was enforced by experienced RUF field commanders who had, largely, been promoted by proven competence in battle. Where the RUF differed from many regular armies, however, was in its lack of a professional non-commissioned officer cadre to maintain the discipline and cohesion of fighting units when not under the direct supervision of their field commanders (Iron 2005). As a result, RUF command was fragile: it relied unduly on a relatively small number of charismatic and experienced commanders; if they were absent or killed then RUF units could disintegrate quickly.

In 1999, a struggle for power within the RUF resulted in considerable intra-factional fighting and took its toll of its leadership: a number were killed or executed. However, many more of the RUF's field commanders were killed during an ill-fated campaign in Guinea in late 2000 and early 2001, conducted at the request of Charles Taylor. The Guinean Army's response, including the cross-border hot pursuit of RUF groups retreating into Sierra Leone, inflicted high levels of attrition on the RUF leadership cadre. A number of its best and most well-known field commanders were killed. This caused a serious weakening of the RUF's military capability as well as reducing the influence of the fighting cadre (who generally supported a continuation of the war) against the political cadre (who sought a political solution) within the RUF.

Although the RUF, when at full strength, was a well-led and competent guerrilla force, we should not overestimate its capabilities. For intelligence, it largely relied on radio intercept of un-encoded Army communications and camouflaged its jungle camps in the visible spectrum only—as a result, it was particularly vulnerable to relatively minor technological enhancements such as secure communications and infra-red surveillance. Once RUF positions were located, they were then vulnerable to indirect fire and air attack. As demonstrated initially by the private military company Executive Outcomes in 1995, the RUF could be defeated by superior organisation—coordinating intelligence, indirect fire, local civil defence forces (usually known as *kamajors*), air and ground manoeuvre into a single operation. The British were again, in 2000 and 2001, able to provide such coordination of assets that were already, in the main, in theatre but had worked separately, not together.

The RUF was Vulnerable To Internal Pressure

Within an insurgent or guerrilla group there is always a tendency towards factionalisation. It is a natural phenomenon—fighting group commanders are often alpha males who tend to resent orders from above and seek the maximum flexibility for their own action; without the strict rank and discipline structures of a regular army, it is easy for separate elements of a guerrilla organisation to go their own way. This tendency is even more marked in a guerrilla group dispersed across a wide jungle area: the senior command finds it difficult to control subordinate commanders who may be operating hundreds of kilometres from the headquarters. It is easy to forget, as a counter-insurgent, that the insurgent has just as many seemingly intractable problems: group cohesion is always an insurgent vulnerability. The RUF was particularly vulnerable to factionalisation without any kind of unifying ideology and it became especially acute after Sankoh was arrested in Nigeria in 1997.

The RUF adopted a number of mechanisms to maintain the loyalty of its dispersed commanders. One of the most effective was to establish a monopoly over ammunition supply: later in the war, all ammunition came from Charles Taylor via the RUF's field commander who could reward subordinate commanders with extra ammunition and starve those whom he considered were not operating in accordance with his wishes. This system generally worked but, when in 1999 the RUF was operating on the outskirts of the Freetown Peninsula, the ammunition supply was at its most tenuous and the system broke down as the RUF split into a six-month contest for the leadership in Foday Sankoh's absence.

The other area of factionalisation within the RUF was between military and political wings. Prior to Sankoh's arrest both wings were united under his leadership, but afterwards he lost control of the military arm of the RUF to his nominal second in command, Sam Bockarie alias 'Mosquito' (SCSL 2003). Sankoh was unable to regain control even after he was rehabilitated by the Lomé Peace Agreement and became a government minister. Only military defeat in Guinea with, concurrently, relentless pressure by the Sierra Leone Army was able to rein in the military wing of the RUF to such an extent that the political wing was able to negotiate peace meaningfully at the review of the Abuja ceasefire agreement in May 2001.

Attempts at Peace

Throughout the eleven-year war, there were numerous attempts to end it. Some were unilateral efforts by the Sierra Leonean Government; but the majority were internationally-brokered peace treaties. Four were particularly important:

1 The Abidjan Peace Accord 1996.
2 The Lomé Peace Agreement 1999.
3 The Abuja Ceasefire Agreement 2001 (Abuja 1).
4 The Review of the Abuja Ceasefire Agreement 2002 (Abuja 2).

None of the peace treaties ended the fighting, apart from Abuja 2. The following section explores the background behind each of the peace efforts and explains why, in the main, they failed.

Abidjan Peace Accord 1996

In May 1995, the Sierra Leone Government was losing the war against the RUF. Rebels were near the outskirts of Freetown and much of the hinterland was controlled by the RUF. The army, where it still occupied bases in the rural area, was under a state of virtual siege. This led to the signing of a contract with private military company Executive Outcomes, a South African-based organisation that largely recruited ex-members of the South African Defence Forces.

Despite never having more than 150 people in Sierra Leone, Executive Outcomes' ability to coordinate government forces was a game changer. By the end of the year, the RUF had been driven out of the diamond centres; had lost many of their jungle bases; and were eking out an existence in the eastern jungles adjacent to the Liberian border. Most of the assets used in this counter-offensive were already available to the Sierra Leone Army: aerial reconnaissance, *kamajors*, helicopters, infantry, Nigerian artillery and Alpha Jets. However, before Executive Outcomes' arrival, their use had not been coordinated into a battle-winning formula.

The military successes achieved with the help of Executive Outcomes led directly, in March 1996, to the first democratic elections held since 1967.

Following a number of national consultative conferences, it had been determined to pursue 'elections before peace'; rather than previous government policy of 'peace before elections' (TRC 2004: Vol. 3A, 220). In other words, democratisation became part of the strategy for ending the war; rather than occurring only once the war had ended.

The RUF tried to impede the elections, and mounted a major effort known as Operation Sabotage Elections (TRC 2004: Vol 3B, 156). Although the practice of amputating hands was already common in the war (Massaquoi 2002: 82, 87), it became widespread during the election period. Operation Sabotage Elections was essentially a RUF campaign of amputation. RUF fighters cut off the thumbs, hands or arms, singly or both sides, of potential voters to prevent them from voting.

Against expectation, perhaps, the elections were widely perceived as fair. Despite the RUF sabotage campaign, 63% of eligible voters cast their votes in the second round. Ahmed Tejan Kabbah, previously a senior official in the UNDP in New York, won 59.5% of the popular vote (African Elections Database undated).

The military successes catalysed by Executive Outcomes also resulted in Foday Sankoh agreeing to negotiate with the government for the first time in the five years of war. Initial discussions were held in Côte d'Ivoire, prior to Kabbah's election, and eventually led to the Abidjan Peace Accord, signed in November 1996. Kabbah maintained the military pressure on the RUF during negotiations, including in October the destruction of The Zogoda, Sankoh's jungle headquarters, in an Executive Outcomes-led operation in which his second-in-command was killed.

The Abidjan Peace Accord specified an immediate ceasefire, the creation of an all-party National Commission for the Consolidation of Peace, a Neutral Monitoring Group consisting of 700 troops, the disarming and reintegration of RUF combatants and the withdrawal of all foreign mercenary groups. As a result, Executive Outcomes' contract was terminated and withdrew in January 1997. Yet, Sankoh was duplicitous: Abidjan permitted the RUF to reorganise and rearm (TRC 2004: Vol 3A, 90).

Just before the Abidjan Peace Accord was signed, Foday Sankoh visited the RUF's main surviving jungle bases in Sierra Leone, to explain to his fighters the need for peace. Publicly, he 'sensitized [the] men on parade the need for giving peace to the people of Sierra Leone', presumably for the benefit of accompanying officials from the International Committee of the Red Cross, who had agreed to provide the helicopter for his jungle journeys. However, when alone with his commanders, he explained that he was in Abidjan to gain a breathing space for the RUF and allow time for a resupply of ammunition (Massaquoi 2002: 50).

Sankoh found help in particular from his early patron Muammar Gaddafi. In a letter to the Libyan ambassador in Ghana, dated 26 June 1996, he wrote:

> I want to thank you and the other brothers at home for the half million United States Dollars (500,000USD) which I received through you for the purchase of needed materials to pursue the military mission.
>
> (TRC 2004: Vol. 3B, 69)

In the same letter, Sankoh went on to request a further US$1 million to 'to purchase twice the listed materials for effective and smooth operation' (TRC 2004: Vol. 3B, 69). He also received financial support from President Blaise Compaoré of Burkino Faso, of about US$50,000 (Massaquoi 2002: 99).

Sankoh sought to buy ammunition from several sources. His primary method was through a pair of Belgian middlemen, who had previously bought RUF-supplied diamonds. Sankoh ordered jungle airstrips to be prepared in Kailahun and Pujehun Districts, so the ammunition could be brought in by air, but the shipment did not arrive until after The Zogoda had been destroyed and the Abidjan Peace Accord signed. He had previously sent an agent to Liberia with US$60,000 to bribe ECOMOG to provide ammunition, but the agent disappeared with the cash. More successful were the RUF's attempts to bribe a Liberian rebel group across the Liberian border. Sankoh sent US$5,000 in cash from Abidjan, sending two of his officers by road through Guinea. With this money, Sam Bockarie (alias 'Mosquito') was able to suborn local rebel forces and obtain sufficient ammunition to permit the RUF to defend Kailahun against attack from government forces. Bockarie had previously been appointed as Sankoh's deputy, after the loss of his predecessor Mohamed Tarawallie in the battle for The Zogoda (Massaquoi 2002: 49–50).

After the Abidjan Agreement was signed, Sankoh again wrote on 4 December 1996 to the Libyan ambassador in Ghana:

> We have signed the Peace Accord on 29 November 1996, just so as to relieve our movement of the enormous pressure from the international community while I will use this opportunity to transact my business in getting our fighting materials freely and easily.
>
> (TRC 2004: Vol. 3B, 26)

The only conclusion that can be reached from this letter and the preceding evidence is that Sankoh never intended to stop fighting until he had captured power. From the moment the government established communications with Sankoh in February 1996, to after the signing of the Abidjan Agreement, Sankoh was using the talks to win a breathing space to re-build the RUF. This is despite the reintroduction of democracy being one of the RUF's initial war aims (RUF 1995).

It is no surprise, therefore, that the Abidjan Peace Accord was never implemented. Sankoh argued about the numbers of the UN Monitoring Group to such an extent that it never deployed; he also delayed nominating RUF members to the Joint Monitoring Group and the Demobilisation and

Resettlement Committee, both mechanisms agreed at Abidjan. Although a few RUF members did emerge from the bush, voluntarily joining the disarmament programme (Penfold 2012), the organisation as a whole made no effort to disarm (Massaquoi, 2002: 50).

In these circumstances, Kabbah's agreement to withdraw Executive Outcomes from Sierra Leone was to have drastic consequences: not only would Executive Outcomes' presence have enforced the RUF to keep its part of the Abidjan Accord, it would also have provided a brake on the Army and, probably, have prevented the May 1997 coup that unseated Kabbah for some eight months.

In an attempt to raise funds to purchase arms during the ceasefire, Foday Sankoh travelled to Lagos on 1 March 1997, where he was promptly arrested by the Nigerian authorities for carrying ammunition. He was to spend more than two years under house arrest in Nigeria (Massaquoi, 2002: 51-6). Nigerian intelligence was perfectly aware of his intention to evade the terms of Abidjan and Sankoh betrayed his naivety in travelling there. Sankoh's withdrawal removed the RUF's strategic leadership and much of the cohesion that bound it together; his new deputy, Sam Bockarie (alias 'Mosquito') was a feared and brutal operational commander, but he lacked Sankoh's personal charisma and strategic outlook. He was also unable to stop the factional in-fighting that affected the movement during the course of 1999 that did much to undermine it from within.

Under Mosquito's violent leadership, the RUF restarted the war against a Sierra Leonean Government that had voluntarily given up its most effective military instrument when it agreed to terminate Executive Outcomes' contract.

The Lomé Peace Agreement 1999

The aftermath of the Abidjan Peace Accord was disastrous for Sierra Leone. Within six months, on 25 May 1997, the army had rebelled, forced President Kabbah to flee to Conakry, renamed itself the Armed Forces Revolutionary Command (AFRC) and invited the RUF to join it in government in Freetown.

The AFRC and the RUF (together known as 'The Junta') was internationally vilified and eventually deposed during a major Nigerian-led intervention by the West African ECOMOG in February 1998 which, together with loyal civil defence forces, recaptured large parts of the country, including some of its main diamond areas. But the RUF was not defeated; a new AFRC faction had been born; and over time both separate guerrilla organisations gained strength in the jungle.

The darkest days for Sierra Leone were about to start. On 6 January 1999, the AFRC bypassed Freetown's defences and sacked much of the capital, causing uncountable deaths and destruction. At the same time, the RUF defeated the Nigerian forces in the east of the country and captured numerous major towns on their march west towards the capital.

Despite a second major intervention of the Nigerian-led ECOMOG, which reasserted government control over Freetown, this was a dire time for President Kabbah's government. The RUF was camped on the capital's outskirts and the Nigerian commitment could not be relied upon indefinitely. Sierra Leone's main international backers, the United Kingdom and United States, were both unwilling to support the government militarily and pressurised President Kabbah to seek a negotiated end to the conflict.

Fortunately for President Kabbah, the RUF was at this time beset by its own civil war as Liberian Denis Mingo (alias 'Superman') unsuccessfully challenged Sam Bockarie's (alias 'Mosquito') leadership of the RUF. The remnants of the AFRC Faction also carved out a role for themselves independent of the RUF in the Occra Hills where they renamed themselves as 'the West Side Boys'.

But this relative stabilisation of the military situation permitted the beginnings of talks which became the second of Sierra Leone's attempts to reach a negotiated end to the Civil War, signed in Lomé, Togo, in July 1999 and brokered by the UN, USA and UK.

The terms of the Lomé Peace Agreement included the appointment of Foday Sankoh as a government minister, withdrawal of ECOMOG (mostly Nigerian) forces and their replacement by a new UN force, UNAMSIL. Unfortunately, UNAMSIL matched neither the Nigerian Army's numbers nor its resolve. Initially, all seemed to go well, as UNAMSIL opened Disarmament, Demobilisation and Reintegration (DDR) camps in Government-controlled areas and demobilised much of the existing Sierra Leone Army. But, in January 2000, when UNAMSIL attempted to expand its deployment into regions dominated by the RUF, conflict immediately arose between guerrillas and peacekeepers:

> There were several serious incidents involving UNAMSIL and former rebel elements or combatants. On 10 January, RUF elements seized a large number of weapons, ammunition and vehicles from a convoy of Guinean troops moving to join UNAMSIL. In two other incidents, numbers of the UNAMSIL Kenyan battalion were ambushed and had to surrender their weapons to ex-Sierra Lone Army combatants in the Occra Hills area on 14 January, and to RUF elements near Makeni on 31 January.
>
> (UNSG, 2000a)

In February, the UN boosted UNAMSIL's mandate and authorised doubling its strength to 11,100 personnel, with the addition of an additional six infantry battalions (UNSC 2000). This was intended to allow the force commander to deploy UNAMSIL into the remaining RUF controlled areas and enable the RUF to begin the DDR process. Unfortunately, the expansion of UNAMSIL's mission took place before the arrival of the troops intended to cover them; as a result, the first two DDR centres in RUF territory were opened on 20 April 2000, when the imbalance of forces between UNAMSIL and the RUF was at its most marked; and when many of the Government's own forces had already been disbanded.

Although there were elements within the RUF that probably did want peace and were prepared to go along with Lomé, the more powerful faction within the group, led by Sam Bockarie in the eastern jungle, had no intention of giving up its arms nor control of its territory. Soon, the RUF responded. On 1 May, they attacked the DDR camp at Makeni and laid siege to the Kenyan Army infantry company and four UN Monitoring Officers (three British, one New Zealander) who were protecting it. A Zambian battalion sent to relieve them was itself ambushed and surrendered *en masse* to the RUF.

News also reached Freetown at about this time that the RUF had seized control of Kambia, a major town in the northern district near the Guinean border. The US Ambassador reported that, without external intervention, the RUF could be in Freetown within a week and the UN peacekeeping mission would collapse (Dorman 2009: 57).

As the situation in Sierra Leone deteriorated, the UN Security Council met in New York on the evening of 4 May 2000. Apart from commending the in-place UN force commander, the Security Council had run out of options to halt the escalating violence: it had no levers left with which to influence the situation on the ground. Immediately following the meeting, the US and French ambassadors, together with the UN Secretary General, Kofi Annan, met the British ambassador and impressed upon him Britain's responsibilities to Sierra Leone as the ex-colonial power: it was now up to the UK to take the lead in rescuing the situation (Dorman 2009: 58).

The Abuja Ceasefire Agreement 10 November 2000 — 'Abuja 1'

The British intervention started on 6 May 2000 and was ultimately successful in creating a level of stability and returning some confidence to a badly-rattled UNAMSIL. Although direct confrontation between British forces and rebels was rare, the British assisted, organised and trained the newly recreated government forces, the Republic of Sierra Leone Armed Forces (RSLAF). British special forces also effectively destroyed the remains of the AFRC, renamed 'the Westside Boys,' in a hostage rescue operation.

On the same day that the first British military elements arrived in Freetown, violence swept across the city. Violent armed mobs attacked the homes of RUF representatives in Freetown, including Foday Sankoh, who had been accepted into Government by the Lomé accord. It is not clear to what extent this attack was orchestrated by anti-RUF elements within the Government, but on that day the political wing of the RUF was effectively decapitated.

Nevertheless, on 10 November 2000, representatives of the RUF and the Government signed a ceasefire agreement in Abuja, Nigeria. The meeting had been convened by a committee established by ECOWAS to facilitate the end of the war. The agreement provided for:

- an immediate ceasefire;
- a monitoring role for UNAMSIL;
- full liberty for the UN to deploy throughout the country;
- unimpeded movement of humanitarian workers;
- the return of UNAMSIL weapons and other equipment seized by the RUF;
- the immediate resumption of the DDR programme;
- a review of implementation of the agreement after 30 days (UNSG 2000b).

Views differ as to why the RUF signed the agreement. One British officer stated:

> Privately the RUF conceded that the British commitment to Sierra Leone was the key factor in their decision to seek a peaceful outcome ….
> (Dent 2002: 23)

Others thought that the RUF was genuinely interested in returning to the Lomé accord, while the Sierra Leonean Government and UNAMSIL wanted the ceasefire to reduce the risk of conflict during the UNAMSIL troop rotation at the end of 2000 (Dorman2009: 120). A more sceptical view, more in line with what subsequently happened, was that the RUF agreed to the ceasefire in order to relieve military pressure on its units as they mounted a major offensive campaign in neighbouring Guinea. Without fear of attack from Sierra Leonean, British or UNAMSIL forces, the RUF could use their territory in northern and eastern Sierra Leone as a safe base from which to mount their attacks into Guinea (Riley 2013).

Many British officers believed that Abuja agreement signalled the effective end of the war (D Richards 2016, interview in London 15 September). Alas, it was not to be that easy. Like every previous attempt at peace in Sierra Leone, unless the agreement was backed by credible threat of force, the RUF simply ignored it while using the time gained to re-arm, re-organise and focus their military effort elsewhere. After Abuja, the RUF's interim leader, Issa Sesay,[1] refused to turn up to or delayed meetings with UNAMSIL to discuss implementation of the agreement. UNAMSIL was not granted freedom of movement in RUF-controlled areas. The only weapons returned during the ceasefire were 11 derelict UN armoured vehicles that had been stripped of all weapons and equipment. No weapons or ammunition were surrendered and no RUF fighters underwent DDR (UNSG 2000b). Nevertheless, the ceasefire held and Sierra Leone enjoyed a largely peaceful Christmas for the first time since 1990.

According to witnesses appearing at the Special Court of Sierra Leone, in July 2000 Charles Taylor ordered the RUF to attack Guinea to oust its then president, Lansana Conte, whom Taylor accused of supporting a rebel group fighting his regime in Liberia. The witnesses claimed that Taylor gave the RUF commander arms and ammunition to support the attack (Sesay 2009).

In late 2000 and early 2001, the RUF conducted a series of major raids in the border area of Guinea, often abducting or recruiting Guineans so to create the illusion that the attacks were conducted by Guinean dissidents. The attacks are estimated to have caused over 1,000 deaths and displaced more than 100,000 Guineans (Global Security undated). For example, the BBC reported on 7 December 2000 that hundreds had been killed in a single major attack on Gueckedou, a Guinean market town near where the borders of Guinea, Liberia and Sierra Leone meet (BBC News 2000).

The Guinean Government was ruthless in its response: in January 2001 it started a campaign of cross-border raids into Sierra Leone to destroy RUF bases. According to Amnesty International, the Guinean Army made no attempt to limit civilian casualties, using multiple launched rocket systems and attack helicopters to attack RUF-occupied towns, causing heavy casualties not just among the RUF but also among civilians. Guinean cross-border attacks stopped on 17 May 2001 after Guineans mistakenly shelled an RUF detachment at a disarmament site (Global Security undated).

The RUF suffered heavily during their ill-fated Guinea campaign and a significant number of their most experienced commanders were killed (Democrat 2001).

Review of the Abuja Ceasefire Agreement 2 May 2001 — 'Abuja 2'

On 2 May, delegations from the RUF and the Government of Sierra Leone met a second time in Abuja to review the implementation of the November 2000 ceasefire agreement. This meeting became known as 'Abuja 2'. It was the first such meeting where the RUF genuinely sought an end to the fighting. The rebels' position had become significantly weaker since the previous November and they realised that, first, their military campaign had been defeated and, second, the political deal on offer was as good as it ever would be.

Ever since Lomé, divisions had existed in the RUF between those who pressed for a political settlement and those who sought continuation of the war. The Sierra Leone Government's decapitation of the RUF's political wing in Freetown in May 2000 had decisively moved the balance of RUF decision-making in favour of the militarists; but many of these were now dead, killed by the Guinean Army during the RUF's foray in Guinea. An opportunity now existed, therefore, for the remnants of the RUF 'peace party' to regain control of the movement.

President Kabbah had been deeply frustrated since Abuja 1 that he was unable to restore Government control over RUF-occupied areas, even when UNAMSIL had deployed there. He was also concerned that he was unable to stop the RUF from using Sierra Leonean territory to attack his neighbour and ally. He felt the Abuja Agreement was being used against him. His solution, with Nigerian and British backing, was to call for a Review of the Agreement, seeking to force the RUF to vacate the north and northwest and permit the RSLAF to assume responsibility for security there. This time, the Government's negotiating position was backed up by the threat of force: if the RUF did not pull out and engage with the DDR process, a strengthened and British-backed RSLAF would feel free to resume hostilities (Riley 2013).

At Abuja 2, the RUF capitulated. They agreed to withdraw from Kambia district, on the Guinean border, and that the RSLAF would deploy there. The withdrawn RUF forces would then undergo DDR (UNSG 2001). The RUF also abandoned demands that the newly formed RSLAF should be also disbanded, once again, and made to go through the DDR process: they recognised that the RSLAF were the established armed forces of the state. Instead, their disarmament would take place in concert with that of the re-mobilised CDF (Sierra Leone Web 2001, 8 May).

The RUF also dropped demands to end the British involvement in Sierra Leone and, unlike Abuja 1, there was no mention of any conditionality to the British remaining. However, Charles Taylor's campaign against the British continued. In April 2001 he criticised President Kabbah for allowing the British to 'practically re-colonise Sierra Leone.' Kabbah bluntly told him to mind his own business (Sierra Leone Web 2001, 12 April).

Abuja 2 immediately revitalised the DDR programme, which was re-launched on 18 May 2001. RSLAF deployments to previously RUF-held border areas continued throughout December 2001 to February 2002 as the official disarmament of districts was declared (Dent 2002: 25). Disarmament formally ended on 17 January 2002. In the period since it restarted on 18 May 2001, 47,076 combatants were disarmed: 19,183 were RUF; 27,695 were CDF; and 198 were AFRC/ex-SLA. 15,840 weapons and 2 million rounds of ammunition were surrendered (UNSG 2002).

By the end of February 2002, the RSLAF were fully deployed around Sierra Leone's territorial borders and the bulk of the country was, for the first time in eleven years, under Government control (Dent 2002: 25). The war was over.

Conclusions

There was no single, simple cause of the civil war in Sierra Leone so it is no surprise that there was no single, simple reason for its end. Military defeats, international pressure and internal contradictions all played a part in the downfall of the RUF and the ending of war.

But the Sierra Leone war ended primarily because the RUF was militarily defeated. Political and economic developments, in particular the democratic election of President Kabbah's government, are sometimes ascribed as being the reasons the war ended, since they largely undermined the original causes of the war. However, the RUF remained remarkably immune to such changes and continued to fight long after most of their original demands had been met, such as the removal of the APC government and the holding of democratic elections.

This is not to say that such political and economic changes in Sierra Leone were unimportant: they contributed significantly to the RUF's defeat. Nor is it true to say that the RUF was defeated purely militarily: it was defeated comprehensively in military, political and economic spheres. It is also not true to ascribe defeat to British military intervention: although important, it was no more so than the interventions by the Nigerian Army, at far greater cost, twice before. In the end it was the Sierra Leoneans themselves who did the bulk of the fighting and freed themselves of the scourge of the RUF.

Despite diplomatic words spoken in New York, the UN and the British in Freetown were both frustrated with each other throughout much of the intervention. The UN could not understand why the UK, as the ex-colonial power, would not contribute its forces to boost UNAMSIL both in numbers and in capability. The British, for their part, were exasperated that UNAMSIL would not take a more aggressive posture towards the RUF, despite a mandate change during the operation that authorised such a position in support of the Sierra Leone government. Both sides misunderstood the different roles that were required: there *was* a place for the impartiality of UNAMSIL to manage the DDR process once the RUF had been defeated. But the RUF was not going to enter the DDR process until they *had* been defeated: this was the job of the Sierra Leone Army with British support. The Sierra Leoneans could not do it by themselves and, at this stage in the war, there were none other than the British who could or would help them. Both roles were required, although neither side understood that at the time; hence the mutual but unnecessary frustration between the British and the UN.

Joseph Opala, an American anthropologist who had lived in Sierra Leone for 17 years, wrote in 2000:

> Soon after the 1996 election, the international community focused all its efforts on negotiating a peace agreement with the RUF. To get a deal, the diplomats reinforced some fictitious notions already rooted in the foreign media -- notions that Sierra Leone was in the grip of a 'civil war,' and the RUF was some kind of political faction.
>
> But far from a faction leader, Foday Sankoh was - and is - a psychopathic killer leading a band of brutalized and confused teenagers. He recruited many of his young henchmen by making them murder their parents, then drugging them with cocaine. Sankoh doesn't want to govern Sierra Leone - he wants to turn it into a criminal enterprise, a base for money laundering and drug smuggling. Sierra Leone does not suffer from civil war, and never has. *It suffers from the rampant banditry that accompanies state collapse.*
>
> [emphasis added] (Opala 2000)

So was the Sierra Leone conflict a civil war? It is common practice to define it so. The CIA World Factbook describes the conflict in Sierra Leone as 'the civil war from 1991 to 2002.' The Wikipedia entry for the war is entitled 'The Sierra Leone Civil War.' The page profiling Sierra Leone on the BBC website also describes the conflict as a 'civil war'.

But if the RUF was not a genuine political entity, had no particular ideology and with no popular legitimacy or support base, it is difficult to see how the conflict could be defined as civil war.

The issue may seem semantic, but it does make a difference as to what should have been the best approach to ending the conflict. If it was genuinely a civil war, then peace-making should include some form of societal reconciliation and power-sharing, such as that mandated at Lomé. If it was an insurgency exploiting a breakdown of order and good government, then peace-making should focus on improving governance while suppressing the symptoms of the breakdown (ie the insurgency). At heart, it comes down to whether the international community was right to have encouraged Kabbah to deal with Sankoh at Abidjan and Lomé, thereby legitimising the RUF; or whether it should have supported Kabbah to improve governance and provided him the resources to deal with the insurgency. The evidence suggests that the latter would have been the better course. This is probably the most important lesson to emerge from Sierra Leone's brutal war.

Note

1 The previous RUF commander in Foday Sankoh's absence, Sam Bockarie (alias 'Mosquito'), was by now working directly for Charles Taylor in Liberia where he was subsequently killed.

References

African Elections Database, undated. Elections in Sierra Leone (tripod.com) [accessed 20 June 2022].

BBC News, 2000. 'Hundreds killed in Guinea attack', *BBC News*, 7 December. http://news.bbc.co.uk/2/hi/africa/1059818.stm [accessed 12 December 2016].

Clapham, C., 2004. Letter to author, 3 July.

Democrat, The, 2001. 'Old Scores Settled … Rambo's men kill Mingo?' *The Democrat*, Freetown, 5 July.

Dent, M.J., 2002. *Sierra Leone Background Brief*, Sierra Leone Joint Support Command D/JSC/COMD/1003, 24 July, in the author's possession.

Dorman, A.M., 2009. *Blair's Successful War: British Military Intervention in Sierra Leone*. Ashgate, Farnham.

Global Security, undated. 'Guinea Conflict', http://www.globalsecurity.org/military/world/war/guinea.htm [accessed 12 December 2016].

Iron, R.M., 2005. *Military Expert Witness Report on the Revolutionary United Front*. Office of the Prosecutor, Special Court for Sierra Leone, 15 April.

Johnson, G., (alias Junior Lion) 2003. Interview with author, 10 October, Freetown.

Kamara, P. 2004. Interview with author, 2 February, Makeni.
Massaquoi, G., circa 2002. *The Conflict*. Unpublished paper in possession of the author.
Massaquoi, G., 2003. Interview with author, 3 November, Freetown.
Opala, J., 2000. 'What the West Failed to See in Sierra Leone,' *The Washington Post*, 14 May.
Penfold, P., 2012. *Atrocities, Diamonds and Diplomacy: the Inside Story of the Conflict in Sierra Leone*. Pen and Sword, Barnsley.
Richards, D., 2016. Interview with author, 15 September, London.
Riley, J.P., 2013. 'Operation SILKMAN', unpublished paper in the author's possession.
RUF, 1995. *Footpaths to Democracy*. Revolutionary United Front. http://www.fas.org/irp/world/para/docs/footpaths.htm [accessed 7 June 2022].
Sesay, A., 2009. 'Taylor Did Not Order the RUF to Attack Guinea,' *International Justice Monitor*, 2 September, available at https://www.ijmonitor.org/2009/09/taylor-did-not-order-the-ruf-to-attack-guinea/ [accessed 17 May 2017].
SCSL, 2003. *Indictment against Sam Bockarie also known as Mosquito also known as Maskita*. Special Court of Sierra Leone, 3 March. SCSL-03-04-I-001.pdf (rscsl.org) [accessed 20 June 2022].
SCSL, 2012. *Sentencing Judgement for Charles Taylor*. Special Court for Sierra Leone, 30 May. http://www.rscsl.org/Documents/Decisions/Taylor/1285/SCSL-03-01-T-1285.pdf [accessed 15 June 2022].
Sierra Leone Web, *News Archive*, available at http://sierra-leone.org/archives.html [accessed on 6 February 2022].
TRC, 2004. *Witness to Truth: Report of the Sierra Leone Truth & Reconciliation Commission*. Sierra Leone Truth and Reconciliation Commission, GPL Press, Accra, Ghana.
UNSC, 2000. *United Nations Security Council Resolution 1289*. United Nations Security Council, New York.
UNSG, 2000a. *Third Report of the Secretary-General of the United Nations Mission in Sierra Leone*. United Nations Secretary General, New York, 7 March.
UNSG, 2000b. *Eighth Report of the Secretary-General of the United Nations Mission in Sierra Leone*. United Nations Secretary General, New York, 15 December.
UNSG, 2001. *Tenth Report of the Secretary-General of the United Nations Mission in Sierra Leone*. United Nations Secretary General, New York, 25 June.
UNSG, 2002. *Thirteenth Report of the Secretary-General of the United Nations Mission in Sierra Leone*. United Nations Secretary General, New York, 14 March.

7 Peace Processes In Aceh and Sri Lanka: A Comparative Assessment

Damien Kingsbury

On 24 December 2004, a massive earthquake in the Indian Ocean off the west coast of northern Sumatra triggered a tsunami that devastated many of the coastlines in the region. The two worst affected areas were the Indonesian province of Aceh in northern Sumatra, and the east and south coasts of Sri Lanka, the eastern coast being predominantly ethnic Tamil.

Both areas had for three decades been involved in separatist conflict. In Aceh, the Free Aceh Movement (Gerakan Aceh Merdeka, or GAM) declared independence from Indonesia in 1976 (Di Tiro 1984), and in late 2004 was fighting the most bitter campaign of the subsequent conflict. In Sri Lanka, the Liberation Tigers of Tamil Eelam (Tamil Tigers, or LTTE), also founded in 1976 (see Swamy 2006: ch 3 for early Tamil militancy), were two years into a ceasefire with the Government of Sri Lanka.

The tsunami provided a catalyst for peace talks over the separatist conflict in Aceh, leading to its eventual resolution in 2005. In the years that immediately followed, Aceh embarked on a democratic peace while Sri Lanka, which had also been affected by the tsunami, returned to full-scale separatist war and eventual destruction of the Tamil Tigers separatist group. This chapter assesses the similarities and differences between the conflicts in Aceh and Sri Lanka, with emphasis on how peace was able to be achieved in one while the other moved away from relative peace towards open conflict and a violent conclusion.

A History of Two Conflicts

Both Sri Lanka and Indonesia have undergone significant internal political tensions over their incorporation of ethnic minorities into their respective states, both of which are unitary. The Acehnese are a distinct ethnic group within Indonesia, having once had their own state, until conquered by Dutch colonialists. Both the Acehnese and Tamils have a distinct territory, culture, language and history of a separate pre-colonial statehood. Their religious practices, however, are different: the Acehnese follow a particularly observant type of Sunni Islam while, in Sri Lanka, most Tamils are Hindu.

DOI: 10.4324/9781003317487-10

In the Indonesian case, its early nationalists opted for the dual strategy of incorporating all Dutch East Indies territories while employing the regional language Bahasa Melayu (market Malay) as the national language (Bahasa Indonesia). The language issue in Indonesia was particularly sensitive given its dozen or so major ethnic groups and more than 300 minor groups. The language of the ethnic majority Javanese is both complex (with five distinct levels) and explicitly hierarchical, which did not suit the state's early egalitarian and inclusive tendencies. In Sri Lanka, the language issue became problematic when in 1956 English was replaced as the country's official language by Sinhalese, with Tamil having no official status, hence effectively excluding Tamils from official communication. This policy was nominally reversed in 1987, although Sri Lanka retains Sinhalese as the functionally dominant language for all ethnic groups.

Claiming self-determination as a separate political entity, a de facto separate Tamil state ('Eelam') existed as a manifestation of Tamil political claims from 2002 until around 2007. A similar claim to separate statehood also existed in the Indonesian province of Aceh, which was resolved with an agreement to grant functional self-government (except in areas reserved to the central government), despite Indonesia also having a unitary constitution. However, its earlier claim to separate state status directly challenged the sovereign cohesion of the pre-existing states and thus led to reciprocal conflict over the last decades.

Indonesia and Sri Lanka share a history of colonialism and, prior to the colonial era, did not have a geo-political unity that has since come to characterize the post-colonial states. Sri Lanka was briefly unified under various warring princes up until the twelfth century but did not enjoy a voluntary political unity and remained disunited after that time (Yogasundram 2006). It was colonized first by Portugal, then Holland and finally the United Kingdom. Large parts of what is now Indonesia, on the other hand, were briefly said to have been unified in the fourteenth century under the somewhat fanciful claims of the Majapahit Empire (Ricklefs 2002: ch2, Cribb 2000: 86–88). Indonesia was colonized by Holland, with Portugal, the United Kingdom and Spain coming to colonize other parts of the archipelago. The pre-colonial existence of both countries demonstrated no prior linguistic or successful, much less voluntary, political unity.

Prior to its colonial incorporation, Aceh had previously in the nineteenth century been recognized as an independent sovereign state by the United Kingdom and the United States (Reid 1969, Reid 1979, Sjamsuddin 1985), a status it still claimed when invaded by the Netherlands in 1873. This invasion was resisted until 1912, although there was sporadic resistance from that time until the Japanese invasion of 1942. Acehnese thereafter directly contributed to the 1945-49 war of independence against the Dutch, assuming they would be granted autonomy within a loosely federated state. However, in 1950, Indonesia was reconstituted as a unitary state and Aceh was subsumed into the province of North Sumatra. In response, by 1953 Aceh had risen in revolt, declaring independence from Jakarta and joining the Darul Islam rebellion with West Java and South Sulawesi. While the Darul Islam rebellion is generally characterized as intended to change the nature of the state rather than to secede from it, Aceh's inclusion reflected a desire to return to at least autonomous status.

The Darul Islam rebellion ended in defeat in 1962, although with a promise of 'special administrative status' for Aceh. However, this was not meaningfully put into practice and deeply undermined after the rise of the New Order government from 1966. Following growing economic exploitation of Aceh by Jakarta, in 1976, a former Darul Islam leader, Hasan di Tiro, declared independence from Indonesia, thus beginning a separatist conflict only resolved in 2005.

In Sri Lanka, as noted by Sitrampalam (undated), the development of Sri Lankan nationalist sentiment during the colonial era was based

> … on the foundations of the society's traditional past. They saw the modern phase of nationalism, not as a novel, essentially different phenomenon, causing a break with the past, but rather as an extension of their past, a rebirth of the old society, its renaissance in a new form.

A culture of national identity that had not yet developed in a civic form led to an attempted hegemony by the majority ethnic Sinhalese over the Tamil and Muslim minorities, which in turn generated conflict as these minorities sought to resist such hegemony (Wickramasinghe 1995).

Sri Lanka's Tamil speaking people had been relatively privileged under British colonial administration, being generally better educated than the Sinhalese majority and employed in government administration. As with other multi-ethnic colonial states that favored an ethnic minority over a majority, such as Rwanda, Uganda and Zimbabwe, when colonialism ended in Sri Lanka in 1949, the ethnic minority lost their privileged position. In 1956, Sri Lanka's administrative language was changed from English to Sinhalese, which alienated many Tamils and led to calls for the establishment of an autonomous Tamil region in the north and east of the country. Tensions and low-level conflict characterized the period from 1956 until the early 1970s.

By 1972, radicalized Tamils resorted to violence in support of their claims, with the precursor group to the LTTE being formed. In 1978, LTTE attacks led to an anti-Tamil Sinhalese riot in which it is believed that up to 3,000 Tamils were killed. This swelled the ranks of the LTTE and, along with covert training from India and later consolidation of militant Tamil groups, led to the formation of a highly developed military organization. By the mid-1980s, the LTTE was engaged in full-scale conflict with the government of Sri Lanka, occupying the Jaffna Peninsula. Indian intervention in 1987 eventually led to conflict between the LTTE and the previously sympathetic Indian army, concluding with the Indian army withdrawing in 1990. Conflict continued, with pauses, from that time until the destruction of the Tamil Tigers in 2009.

Similarities and Differences Between Aceh and Eelam

While all states are different—such difference is implied in their geo-institutional separation—and all conflicts are specific to their own causes and circumstances, some states share commonalities and some conflicts do find points of common reference. Eelam and Aceh share a number of commonalities, specifically that they were both parts of colonial unities that did not reflect pre-colonial political conditions, that both existed within a unitary post-colonial structure, both had claims to pre-existing national identity as a state and which continued to manifest as post-colonial national identities, and that both had been engaged in a war for self-determination.

Both movements also adopted the interconnected universalist claims to and rhetoric of self-determination, democratization (variously defined or misunderstood) and human rights. In practical terms, the claims to nationalist self-determination by both provided a more significant similarity between them than their different religious orientations might have implied. Following this, both movements also appeared to be driven by an adherence to a sometimes-fatalistic attitude towards dying for one's cause.

Yet the similarities between the Aceh and Eelam conflicts are much tested by their differences. The first main difference between the two is that the Aceh conflict, while reflecting more conventional claims to nationalism, was motivated by Islamic principles. The Eelam conflict was similarly nationalist, but not clearly inspired by any particular religious preference, even though most Tamils are Hindus (a small minority are Christian). A more significant ideological difference between the two organizations was that while GAM practised communal living in its liberated zones, the natural tendency of its senior members both in Aceh and abroad was to gravitate towards small business. Indeed, GAM's original leadership largely comprised of small businessmen, traders and professionals, and their rebellion can in part be seen as opposed to the imposed centralized corruption of Jakarta. However, the conflict quickly produced depredations against Acehnese, which became its principal motivating cause. Beyond a loose orientation towards small business, GAM completely lacked an official ideology, except for a relatively late and, for some, superficial conversion to democracy, and was otherwise generally politically moderate. While technically hierarchical, its organizational structure was relatively flat and decentralized, with field commanders having a high level of operational autonomy, and with the reporting process to the political leadership in Stockholm being primarily advisory.

The LTTE, on the other hand, had its roots in the student movements of the 1970s, was revolutionary and tended towards state socialism, if blended with a high proportion of small business activity and an in-principle acceptance of free market capitalism (Stokke 2006). The LTTE was a distinctly hierarchical and highly centralized organization which, according to the head of the LTTE Peace Secretariat, Puledevan (Kilinochche, May 11 2006), precluded democratic processes on the grounds that it could not afford internal dissent in times of war.

GAM officially endorsed a 'democratic' platform in its 'Stavanger Declaration' (GAM 2002) but had not worked out a mechanism for a democratic internal structure until a meeting in Stockholm in February 2006 (in which the author participated). GAM 'prime minister' Malik Mahmud thereafter subverted the agreement on the democratic internal selection of candidates for elections, in May 2006 moving to appoint candidates in opposition to candidates democratically selected by the organization just days before, precipitating a post-war split within the GAM and the creation of two distinct political parties, Partai Aceh and Partai Nasional Aceh.

A further significant difference between the two organizations was that the LTTE was widely and increasingly proscribed as a terrorist organization, whereas GAM was not identified as such outside Indonesia. The implication of this was that being associated or working with the LTTE was a criminal act in many parts of the world, whereas being associated or working with GAM was not. This in turn had implications for travel, fund-raising activities, the distribution of aid to sympathetic in-country organizations and a host of security-related matters. This 'terrorist' distinction characterized different approaches to conflict by GAM and the LTTE. GAM was careful not to engage in activities that could see it labeled as a terrorist organization, in particular the indiscriminate killing of civilians, the use of bombing to create a sense of terror, or operations outside its specific field of concern. The LTTE, by comparison, has engaged in indiscriminate attacks, has been credited with inventing tactical suicide bombings, and often engaged in attacks well outside its principle geographic area of concern. Its general justification has been that such attacks were a legitimate response to government attacks within its claimed homeland and that it could not afford to limit its operational capacity by concerns about how it was externally viewed.

More importantly, and with serious implications for peace processes, GAM had significant guerrilla capacity but, with a potential force ratio of around 1:40, or approximately 5,000 against 200,000 (the figure of 3,000 fighters in the MOU 2005 was intentionally understated by GAM), was never enough to militarily defeat the Indonesian military (Tentara Nasional Indonesia, or TNI), other than in occasional local battles. Indeed, as GAM regional commander told the author in Aceh, in 2003, and based on firsthand observation, GAM was generally only lightly armed and struggled to meet its own requirements for armaments and munitions. The head of the first intake told the author in Stockholm in 2005 that its core fighters had been trained as specialist combatants in Libya, but most of its guerrillas were less well trained and equipped. But based on conversations held in KIlinochchi in 2006 with several LTTE cadre, the LTTE could field a conventional army of up to 25,000, was well trained, heavily armed, had a functional naval unit (Sea Tigers) and even a small air wing, which it used to symbolic effect in an air strike on a military air base in Colombo in late March 2007. By comparison, in the early 2000s the Sri Lanka Armed Forces had perhaps 158,000 personnel whose training was generally no better than that of the LTTE, which initially lacked motivation and, while better equipped with conventional weapons, ships and planes, until the final three years of the war did not enjoy a consistent advantage in conventional combat.

Related to the capacity or otherwise to field conventional military forces was the state capacity of the respective organizations. While neither GAM not the LTTE could claim to represent a state that was externally regarded as legitimate, both did administer significant areas of territory, or 'liberated zones', and this thus implied some state capacity. GAM's 'state capacity' was limited to local administrative structures based on traditional village models (Kingsbury 2007), limited educational facilities and rudimentary health facilities. When GAM fighters needed serious medical treatment, in most cases they would pay a local doctor or, a senior GAM organizer in Kuala Lumpur told the author, in 2001 and 2002, transfer to Malaysia where there was a considerable Acehnese community. Education, beyond most basic levels, was conducted through government funded schools and institutions, both inside and outside liberated zones.

Compared to this most basic level of organizational capacity, the LTTE virtually ran a state within a state and its intention was, rather than to wait for a negotiated settlement, it would establish a de facto independent state. The status of this 'state' was enhanced by the 2002 ceasefire agreement and technical control lines which quickly morphed into 'borders' which, based on firsthand observation at the Vanni technical control line (9 May 2006), was complete with immigration checks, visas and customs and tax. The 'capital' of Eelam, Kilinochchi, was the LTTE's administrative center and hosted a range of institutions such as police (including traffic police with speed detection radars—speed was limited to 50 km/h within towns and 80 km/h outside), three levels of courts and a detailed legal code (LTTE 2004). Other signs of administration included departments of works, public transport and fuel, hospitals, asylums, clinics, children's homes, rehabilitation centers, primary and secondary schools and the construction of a proposed university.

Elements in Favor of Resolution in Aceh

The resolution of conflict requires a genuine desire to achieve peace, the intention to do so, and having the capacity to control one's forces to ensure that peace can be implemented and sustained. When considering the option of a negotiated peace in Aceh, GAM's political leadership in Stockholm was asked by the author (27–30 October 2004) whether it was prepared to accept anything less than full independence. This was predicated upon the first principal question of what independence was intended to achieve. Once having established what purpose independence was intended to serve, it was then possible to ask whether that purpose could be achieved by means other than independence.

Similarly, during the negotiations process, the author put the question to the head of the Indonesian negotiation team, Justice Minister Hamid Awaluddin, in Jakarta on 23 March 2005, that if the integrity of the state was its principal goal, could this be achieved by means other than imposing a constrained unitary state, which in this case only allowed political parties that had wide representation to participate in elections, precluding locally based parties. This then went back to questions about the principal purpose of the state, and whether such constraints were necessary to maintain state unity. Finally, GAM was a significantly united organization and while it could not claim demonstrated widespread representation within Aceh, it could claim a monopoly on the means of violence by pro-self-determination forces.

There were also strategic considerations by both sides to be taken into account. The strategic reality for GAM in 2004 was that, as an organization, it had suffered serious reversals under a heavy renewed military campaign launched in May 2003. While GAM's numbers were not significantly depleted, its access to sources of food, medicine and ammunition was constrained, damaging the morale of its fighters and their capacity to sustain their military claims, much less hold territory. Similarly, with a population of a little over four million, Aceh comprised less than 2% of the population of Indonesia (at that time around 230 million). However, the strategic situation was not entirely in the government's favor. The TNI had shown that while it was capable of inflicting suffering on GAM, it did so primarily by inflicting suffering on the people of Aceh. This had the effect of driving more recruits into the arms of GAM, strengthening it at the time it was under heaviest attack. Similarly, the TNI had shown, over almost thirty years of conflict, that its military approach was unable to end the guerrilla campaign. While it had damaged GAM, the organization was by no means defeated, and had demonstrated in the past great capacity to come back strongly from previous reversals. In this sense, while GAM survived it was in fact winning.

Having noted GAM's capacity to survive and to regenerate, by 2004 it had also become clear that GAM simply did not have the military capacity to win independence through military means, and that there was no international support for Aceh's independence, which precluded any possible international settlement in that direction (such as in East Timor). Balanced against this, the Indonesian government also needed to exercise control over the TNI. The TNI had long been a political and economic power in its own right, was engaged in widespread corrupt and illegal activity, and was only nominally under civil authority (Mietzner 2003, Kingsbury 2003: ch 5). Statements made by President Yudhoyono indicated that civil authority over the TNI was one of the key objectives of his presidency (Laksamana 2004). To achieve that, he needed to separate the TNI from its sources of income, in particular in Aceh, and to reduce the TNI's claim to being a self-appointed guardian of the state. Added to this was that, set against a tight budgetary environment, the cost of the Aceh campaign was problematic for the central government. Reliable figures for the cost of the campaign were unavailable, but it was known that the campaign had a budget for six months, which it exceeded when the campaign was not successful in that time. There were also unconfirmed accounts of the TNI levying other government departments to bolster its funds for Aceh.

The conflict also had the effect of scaring off potential foreign investors, especially in hydrocarbon industries such as those in Aceh. This meant that Indonesia had gone from being a net oil exporter to becoming a net oil importer, at a time of world record high oil prices. Further, the lack of foreign direct investment approximated the gap between Indonesia's economic growth, at around 4.5%, and population growth, meaning that per capita GDP was declining rather than growing which, based on relatively steady incomes, meant increased unemployment. Hence Yudhoyono needed a resolution to the Aceh conflict for a number of pressing domestic reasons, as well as to show that Indonesia could resolve its internal problems and make a stronger claim to being a nation as well as a state. In this, Yudhoyono could claim a popular mandate as Indonesia's first directly elected president and, through strategic alliances (including his vice-president, Jusuf Kalla, who was also head of Golkar, the largest party in the legislature), could rely on majority support within the Indonesian legislature.

GAM similarly faced economic difficulties, although of a different type. Apart from having reduced access to supplies, the local and expatriate community also had difficulty in funding GAM, while GAM's levying of a *pajak nanggroe* (state tax) was, GAM's senior intelligence operative told the author (Stockholm, May 2005), resented by some and otherwise increasingly difficult to collect. In all, GAM's economic circumstances were considerably reduced, which made continuing the conflict difficult, although far from impossible. Finally, the tsunami on 24 December 2004 that devastated most of low-lying Aceh, left more than 180,000 people dead or missing and many more homeless provided significant impetus to a decision by GAM made two days prior to accept an invitation to attend talks in Helsinki.

In the first instance, the tsunami opened the province, which had been closed to outsiders since May 2003, allowing in large numbers of media as well as foreign militaries (as aid suppliers) and large numbers of aid organizations. This had the effect of highlighting both the conflict (GAM declared a unilateral ceasefire immediately following the tsunami, although the TNI continued its operations) and the necessity of ending it to allow relief work to proceed unhindered. While the TNI was initially reluctant to allow in foreigners, the extent of the disaster was such that the Indonesian government was unable to cope and quickly accepted foreign offers. Similarly, while a TNI function was supposed to be development and emergency relief, it was overwhelmed by the scope of the disaster and, according to some (first person) aid agency reports (Aspinall 2005b), even slowed down the shipment of supplies to disaster areas and imposed 'taxes' on goods being shipped (consistent with its common 'revenue' raising methods).

If both sides wished to see an end to the conflict, they had until this time been trapped by their own absolutist rhetoric and the intensity of the conflict. The tsunami acted as a circuit-breaker to these impediments and allowed them to assume the high moral ground in seeking peace (despite the TNI continuing its offensive campaign throughout the peace process). On the part of GAM, too, there were also frequent references by GAM negotiators in conversation with the author in Helsinki during the negation process, that 'the people of Aceh have suffered enough.' Finally, once the attention of the international community was turned to Aceh, not least through media access, there was a view that both sides should compromise to reach a negotiated settlement (Kingsbury 2006: 34).

In peace processes there are usually elements that militate against a resolution, which might undermine the process itself or could undermine any resolution that is achieved. In Aceh, the main problem following the signing of the peace agreement in August 2005 was that the enabling legislation was passed four months after the agreed date, in July 2006, and that it compromised a number of elements, in particular removing the Acehnese legislature's power of veto on state legislation concerning Aceh, the method of allocation of income from natural resources, and other issues of central control (Tempo 2006a). However, the GAM negotiating team told the author in Helsinki on 16 July 2005 and 13 August 2005, that sufficient of the original agreement remained, in particular a freeing of local political opportunity, to allow GAM to continue with the peace process. While the incomplete implementation of the peace agreement (MOU 2005) was problematic, more so was the continuing presence of what were claimed by GAM to be an unnecessarily high number of TNI and police. 14,700 troops and 9,100 police were to remain stationed in Aceh (MOU 2005: 4.5), which was around double the usual number for a military command area or a province. Further, both TNI and police were reluctant to accept the changed circumstances and both continued to engage in illegal activity in the province, to the detriment of local people. Associated with this was a lingering sense of bureaucratic-authoritarianism on the part of the Indonesian government, if not on the part of the executive branch, then certainly in elements of the legislature and the bureaucracy (Tempo 2006b). Finally, while the MOU did allow Aceh some degree of political autonomy, questions remained as to how adequately this addressed the underlying sense of Acehnese national identity, and the relationship between such an identity, self-determination and claims for an independent state (see Aspinall 2005a). This did not undermine the peace in Aceh, which was cemented after Aceh's administrative elections in December 2006, in which (unofficial) GAM and GAM-supported candidates won about 80% of all positions, including that of governor and vice-governor. It did, however, provide a continuing backdrop of nationalist assertion that has since tested the relationship between Aceh and Jakarta.

Elements Militating against Resolution in Eelam

By way of comparison, in Sri Lanka the issues of interest, intention and capacity were more complex. Indonesia was constructed of dozens of significant minorities (along with hundreds of small minorities and one majority of a little over half of the total) and hence incorporating minority concerns (even if that has been incomplete) is critical to state success. In Sri Lanka, however, there is one large majority—Sinhalese (about 70%)—and two significant minorities, one being Muslim (about 7%) and the other Tamil (about 15%) (Gunasingham 2005: 12–32). Within the Sinhalese majority there was a tendency towards a national chauvinism (De Votta 2004), underpinned by the dominant religion of Theravada Buddhism, which constitutionally occupies the state's 'foremost place' (Sri Lanka Constitution II.9). While some Buddhist clergy and leaders have worked for peace, others from its monastic order have been religiously assertive and, in some cases, combative. In this, the view is that the country of Sri Lanka is a global repository of Buddhist values and hence must be maintained as wholly Buddhist. Within this world view, Hindu Tamils and Muslims (also mostly Tamil) are an unwanted intrusion and should preferably convert to Buddhism or otherwise subsume their sense of difference. Although geographic proximity worked against the LTTE—with the exception of the Jaffna Peninsula, most Tamil areas are easily accessible from the rest of Sri Lanka—the relative size of the Tamil population and its concentration, especially in the north and to a lesser extent the east, appeared to create a viable state alternative to the Sinhalese dominated Sri Lankan state. Similarly, until its demise, the LTTE had a monopoly on the use of violence within the area it controlled, and hence imposed unity upon the claims to Tamil self-determination.

Finally, while the 2004 tsunami had a serious impact on Sri Lanka, with some 30,000 being killed, it did not draw in foreigners relative to local population to the extent of Aceh, nor did it act as a catalyst for peace. Rather, the tsunami and the aid flows associated with its relief acted to further entrench divisions within the state, especially over the allocation and control of the flow of aid. The LTTE demanded the right to allocate aid within the area it controlled, while the government insisted that aid was a whole of state responsibility. In reality, aid represented a significant financial contribution and whoever controlled that stood to strengthen their own position vis-à-vis the other.

The impediments to a negotiated resolution in Sri Lanka were thus more profound. From the perspective of senior members of the LTTE international branch, in particular in interviews by the author with the head of the LTTE Peace Secretariat, Pulidevan (10 and 12 May 2006, Kilinochchi), the head of the LTTE International secretariat, Castro (nom de guerre) (11 May 2006, Kilinochchi) and the political head of the LTTE, Suppiah Paramu Tamilchelvan (13 May 2006, Kilinochchi), at a time in which there was also a marked breakdown in the 2002 'Cease Fire Agreement', there was a clear lack of trust in the intentions of the Sri Lanka government to genuinely pursue peace. This was supported by escalating attacks by Government forces over 2006, in particular from May. These attacks were formalized by the announcement at the beginning of January that the Government of Sri Lanka intended to seek a military solution to the LTTE issue (Balachandran 2007). Moreover, while Indonesia had been moving towards an increasingly democratic system, Sri Lanka appeared to be moving towards a type of bureaucratic authoritarianism. While an outcome that promoted democratization and political devolution could work in Aceh, it appeared to have little hope in Sri Lanka, due to the extensive and, for a time, seemingly autocratic powers of the executive president who was, almost by default, an ethnic Sinhalese.

Related to the LTTE's state capacity was its military capacity, which challenged that of the Sri Lanka military in various conventional military operations. That is to say, while the LTTE could not have won the war outright against the Sri Lanka government, until around 2006–2007 it did appear to have the capacity to take and hold significant territories. If the LTTE felt pressure to that time, then it was pressure of a similar type felt by the Sri Lanka Government and its military. In this, the relatively even matching of the two sides, if not in numbers then in capacities, until around 2006–2007 did not introduce the disequilibrium that pushed one side into a position of seeking a negotiated settlement.

While there were expressions of interest within the Sri Lankan government in a negotiated settlement, approaching genuine negotiations—those that could fundamentally alter the nature of the Sri Lankan state—was exceptionally difficult in Colombo. Sri Lanka's legislature was controlled by a series of coalitions around one of two major parties. Although the two major, almost exclusively Sinhalese, parties (United National Party and Sri Lanka Freedom Party) then dominated, neither was able to command an absolute majority in its own right and hence both are vulnerable to losing the legislative majority. Further, both had tended to try to 'outbid' each other with nationalist Sinhalese rhetoric. Any government that appeared to compromise with the LTTE thereby opened itself to attack by opposition parties, or indeed by minor coalition partners, and hence was likely to be undermined in elections or in office. This then acted as a structural political impediment to the legislature pursuing or accepting a negotiated settlement. Similarly, while Sri Lanka had an executive president, he or she still relied on the legislature for the passage of enabling legislation and may similarly be undermined in office, assuming a prior intention to compromise.

Sri Lanka shifted from a parliamentary to a presidential-led system (akin to the French model) in 1978. Since then, executive authority has been increasingly located in the hands of the president, who appoints ministers as well as the Supreme Court judges who decide on constitutional issues. The role of the prime minister has since become increasingly ceremonial. The president appoints, among others, the defence minister, who was from late 2005 was army brigadier-general and brother of the president. Parliamentary politics continued in Sri Lanka, although the use of army-linked death squads, torture and denial of *habeus corpus* contributed to an increasingly constrained political climate (USAID 2006).

From December 2005, the Sri Lanka government took a significantly more belligerent attitude towards the LTTE, escalating its attacks via proxy 'militia' (including the so-called Karuna faction) and engaging in direct kidnappings, murders and assaults. Since the ceasefire of 2002, both sides had also upgraded and stockpiled weapons ahead of a new round of fighting. By 2006, the Sri Lanka government sought to again control all of the territory of the state prior to negotiations. Similarly, the LTTE sought to expand its territorial control, especially to areas claimed as Tamil homeland, to be able to negotiate its claims from a position of relative strength. It was from this point that both sides resumed and escalated the conflict, with both seeking an effective military victory, or to be able to negotiate from a position of strength.

Conclusions

Both the conflicts in Aceh and Sri Lanka showed that where the state was perceived to have failed in its civic responsibility towards an ethnically distinct and geographically coherent minority, that minority could retreat to or create more localized conceptions of nation. In turn, this 'nation' could seek territorial independence, that is a state, through which to represent its political claims. This then created a separatist agenda which, as a challenge to the pre-existing state's sovereign authority, led to conflict. Resolution to such conflict in part rested on partially acknowledging the evolved legitimacy of such a separatist claim and, by way of compromise, allowing for some of that claim to be manifested in practice as autonomy or federation. Such compromise implicitly recognized the government's obligation to manifest its civic responsibilities not just as a sovereign authority but as a civic guarantor. In cases where the state has a poor civic record, external monitors could help ensure compliance with such a process.

The Aceh conflict was resolved by the government of Indonesia agreeing to allocate to the people of Aceh a degree of genuine (as opposed to the previously offered nominal) political autonomy, within a less constrained unitary structure. In particular, the Indonesian government agreed to allow for the creation of local political parties and accepted local independent candidates for political office in Aceh. These potentially gave substance to other claims of autonomy, by advocating and representing local political wishes as opposed to being a branch office of a Jakarta-based party. This was the key to achieving peace in Aceh, and while other elements of the MOU were important, the whole agreement would have succeeded or failed on this single issue. Indeed, it was the last matter to be negotiated by the two teams at the peace talks and prior to its conclusion appeared to be the one issue that would result in the negotiations failing.

There were two further elements which helped secure the Aceh peace. The first was international promises of support and supervision of the peace process, in particular through the European Union's Aceh Monitoring Mission (AMM). While there were numerous criticisms of the performance of the AMM, it did generally fulfill its main functions, not least of which was supervising GAM's disarmament. The second was the relative success of Indonesia's shift from procedural to a more substantive democracy, the scope within this for local democratic outcomes, and the dedication of then recently elected President Yudhoyono to what might be described as a civic national project, as opposed to the imposed 'nationalism' of his predecessors.

By the end of 2006, when the people of Aceh were going to the polls for the first time in three decades in a state of peace and, for the first time ever, able to elect a locally constituted candidate, the ceasefire that had endured in Sri Lanka from 2002 no longer functionally existed and the country was plunged back into an undeclared war. The LTTE's offer of a federal solution, based on unified northern and eastern provinces, was withdrawn at this time as it again called for a completely independent state. Where the AMM had been relatively successful in its mission, Sri Lanka had the experience of the intervention of the Indian Peace Keeping Force, which resulted in it battling the LTTE and eventually and ignominiously withdrawing. The very much smaller and unarmed Sri Lanka Monitoring Mission (SLMM), established to oversee the 2002 Cease Fire Agreement, was unable to prevent Sri Lanka's slide back into conflict. While all such agreements require the active support of all combatant parties, the SLMM lacked the resources or political capacity to impose its will on the combatant parties.

In a practical sense, the LTTE's creation of a state within a state was a concrete step towards the realization of its aims, and its functional acceptance by the Sri Lanka government via the 2002 ceasefire agreement confirmed there was a starting point for compromise. But perhaps, more than anything, the events of 2006 and into 2007 demonstrated that there was little taste for compromise in Sri Lanka. The depredations by both sides entrenched an abiding sense of mutual antipathy which was perhaps the biggest hurdle for either combatant to overcome. Both sides appeared quite willing to shed much more blood before they might, exhausted and depleted, have returned to the negotiating table. As it turned out, negotiations were no longer being pursued.

Arguably, too, the LTTE's creation of a 'state to be' was part of its downfall, the last and most significant of the LTTE's mistakes. The LTTE had alienated its initial biggest supporter, India, in 1991 by assassinating Rajiv Gandhi, in retaliation for ordering in 'peace keeping troops' to Sri Lanka in 1987. Then, in 2005, the LTTE insisted that Tamils boycott the presidential elections, helping ensure a narrow win by hardliner Mahinda Rajapaksa, rather than Ranil Wickremesinghe who was more amenable to a negotiated settlement. Rather than negotiate for territory it already controlled within a functionally federated framework, the LTTE said that it would only negotiate after it had recaptured Trincomalee, which was to be Eelam's new capital (it never came close to recapturing that lost territory). Finally, where GAM did not have a state and was able to continue asymmetric and malleable guerrilla operations against a much larger force, the LTTE adopted the pretentions of a state and attempted to field a conventional army in conventional, fixed battle which, as a nascent proto-state, it did not have sufficient capacity to sustain.

The Return To War

In November 2005, Rajapaksa was elected to the presidency on a clear policy of militarily defeating the LTTE but with a slim margin of just 1.8%, which many attributed to the LTTE stopping or 'discouraging' people under its control from voting, in this case for the opposing candidate. Following Rajapakse's election and the appointment of his brother Gotabhaya Rajapaksa as Defense Minister, the Sri Lanka government became significantly more aggressive towards the LTTE, escalating its attacks via proxy militias, engaging in direct kidnappings, murders and assaults, and eventually full-scale military confrontation. The government's policy of having the LTTE listed as a terrorist organization by other governments became increasingly successful and its effect was to limit fundraising for the LTTE which denied it an important source of income. The ceasefire agreement was more or less respected until early 2006 but from early in the year it began to fall apart, with both sides escalating attacks against the other. By mid-year, following full-scale government attacks in both the north and east, what was referred to by Tamils as Eelam War IV had begun. This return to war demonstrated that there was little scope for a negotiated settlement and that both sides were belligerents. Only significant external intervention at this time could have halted the continued slide into full-scale war and the atrocities that accompanied it.

Despite initial signs of LTTE strength, in 2006 the Sri Lanka army conducted a series of successful campaigns. The first was to push the LTTE out of the eastern provinces, including an LTTE artillery site near Trincomalee, then the capture or destruction of LTTE-owned ships carrying weapons and supplies. The LTTE resumed its bombing campaign, with one suicide bombing at Habarana killing around 100 sailors in the single deadliest attack of the war. By 2007, the army assaulted the LTTE's northern bases, initially targeting the LTTE's leadership with air strikes. In November 2007, the LTTE's political leader, Thamilchevan and five other senior leaders were killed in what was described by the LTTE as a precision air strike. Despite these setbacks, LTTE forces on the ground put up a stiff defence against continuing army assaults. After a number of setbacks, from August 2008 the SLA eventually advanced, taking Mannar in the west and then assaulting the LTTE's 'capital' of Kilinochchi from three sides in November 2008, capturing it in January 2009. With the loss of their 'capital', the LTTE, civilians under its protection and others retreated to Mullaitivu where they underwent a sustained air attack, including against civilians sheltering with or being held by the LTTE. Again overwhelmed, by the end of January 2009 the remaining LTTE forces and civilians retreated further. It was clear from this time that a humanitarian catastrophe was unfolding but, due to vetoes by two of its permanent members, China and Russia, the UN Security Council was helpless to act despite a resolution to try to stop the slaughter.

In the end, the LTTE and a civilian based held a narrow, 14-kilometre strip of beach near Mullaitivu. With some 20,000 civilians killed in the government's declared 'no fire zone' in last weeks of the war, at the almost inconceivable rate of about 1,000 a day, the world stood by and watched this humanitarian disaster unfold. A total of around 40,000, predominantly civilians, were killed in the final months, adding to the 100,000 or so that had been killed previously.

The main reason why the international community was not able to intervene in the final stages of the anti-LTTE war was because it was both blocked in the UN Security Council by China and Russia, and it was blocked on the ground by China and India both actively supporting the government of Sri Lanka—India with intelligence and China with weapons and funds. China's involvement had compelled India to support the government of Sri Lanka, to ensure that its influence was not neutralized in favour of its regional competitor.

> Sri Lanka was able to vanquish the Tigers ... due to India's tacit support, which included providing training and logistical support such as naval surveillance to cut off Tigers' weapons supply. There were two motives: India wanted to be rid of the LTTE and India could not sit back and see its regional influence overwhelmed by its main strategic competitor. Against the combined military might of Sri Lanka, India and China, the LTTE was hopelessly outclassed.
>
> (Pratap 2009)

References

Aspinall, E., 2005a. 'The Helsinki Agreement: A More Promising basis for Peace in Aceh?' *Policy Studies No 20, East-West Center.* International Crisis Group, Washington.

Aspinall, E., 2005b. 'The Americans brought equipment, the Germans brought medical supplies, the Australians brought clean water but the TNI only brought guns', *Ed Aspinall, Center for Southeast Asian Studies seminar on 'Post-tsunami Aceh'.* Monash University, 27 April.

Balachandran, B., 2007. 'Lankan army hopes to clear East by March', *Hindustan Times.* New Delhi, 3 January, a newspaper.

Cribb, R., 2000. *Historical Atlas of Indonesia.* Curzon, London.

De Votta, N., 2004. 'Chauvinism and Ethnic Conflict in Sri Lanka', *BASAS Annual Workshop*, University of Bristol, 13 November.

Di Tiro, H., 1984. *The Price of Freedom: The Unfinished Diary by Hasan di Tiro.* The Open Press, Ontario.

GAM, 2002. 'Stavanger Declaration', *Aceh-Sumatra National Liberation Front.* Stavanger, Norway, 21 July.

Gunasingham, M., 2005. *Primary Sources for History of the Sri Lanka Tamils: World-Wide Search.* MV Publications, Chullora, NSW.

Kingsbury, D., 2003. *Power Politics and the Indonesian Military.* Routledge-Curzon, London.

Kingsbury, D., 2006. *Peace in Aceh: A Personal Account of the Aceh Peace Process.* Equinox, Jakarta.

Kingsbury, D., 2007. 'Islam, Democratisation and the Free Aceh Movement', *Journal of Contemporary Asia,* vol 37, no 2, pp. 166–189.

Laksamana, 2004. 'TNI Polemic Shows Military's Political Clout', November 6. Accessed at http://www.kabar-irian. com/pipermail/kabar-indonesia/2004-November/000895.html 2 February 2007.

LTTE, 2004. *The Judicial System of Tamil Eelam: Structure Function and Duties of Officials in Charge.* ANBU Printing Press, Kilinochchi.

Mietzner, M., 2003. 'Business as Usual? The Indonesian Armed Forces and Local Politics in the Post-Soeharto Era', in Aspinall, E. and Fealy, G. eds., *Local Power and Politics in Indonesia: Democratisation and Decentralisation,* ISEAS, Singapore.

MOU, 2005. *Memorandum of Understanding Between the Government of the Republic of Indonesia and the Free Aceh Movement,* Helsinki, 15 August. The Crisis Management Initiative.

Pratap, A. 2009. 'Lessons from the Tiger defeat', *The Week.* 31 May. A newspaper.

Ricklefs, M., 2002 *A History of Modern Indonesia Since C. 1200.* Stanford University Press, Stanford.

Reid, A., 1969, *The Contest for North Sumatra: Atjeh, the Netherlands and Britain, 1858-1898.* Oxford University Press, Kuala Lumpur.

Reid, A., 1979. *The Blood of the People: Revolution and the End of Traditional Rule in Northern Sumatra.* Oxford University Press, Kuala Lumpur.

Sitrampalam, S., undated. 'Nationalism, Historiography and Archeology in Sri Lanka—A Perspective unpublished paper'. University of Jaffna, Thirunelvely.

Sjamsuddin, N., 1985. *The Republican Revolt: A Study of the Acehnese Rebellion.* Institute of Southeast Asian Studies, Singapore.

Stokke, C., 2006. 'Tamil Eelam—a De Facto State: Building the Tamil Eelam State: emerging state institutions and forms of governance in LTTE-controlled areas in Sri Lanka', *Third World Quarterly,* vol. 27, no. 6, pp. 1021–1040.

Swamy, N., 2006. *Tigers of Lanka: From Boys to Guerrillas 7th ed.* Vijitha Yapa Publications, Colombo.

Tempo, 2006a. 'Bones of Contention', *Tempo* No. 46/VI July 18-24.

Tempo, 2006b. 'Passed … Imperfect', *Tempo* No. 46/VI July 18-24.

USAID, 2006. *Mission in Sri Lanka, Democracy and Governance Program, Democracy and Governance Program Initiative.* United States Assistance for International Development program, Colombo.

Wickramasinghe, N., 1995. *Ethnic Politics in Colonial Sri Lanka 1927-47.* Vikas Publishing House Pvt Ltd, New Delhi.

Yogasundram, N., 2006. *A Comprehensive History of Sri Lanka: From Prehistory to Tsunami.* Vijitha Yapa Publications, Colombo.

8 The Case of Afghanistan: How Wars End[1]

David Kilcullen

A Slow Deterioration

Although the collapse of 2021 seemed to come quickly at the end, its roots stretch back at least to 2014. When the International Security Assistance Force (ISAF) withdrew from Afghanistan at the end of that year, coalition forces transitioned from counterinsurgency (COIN) to security force assistance (SFA). Under the COIN strategy—exemplified by President Obama's 'surge'—international forces carried the main combat burden. The goal was not to completely defeat the Taliban but rather to weaken the insurgents to the point where Afghan forces could take the lead in countering them, with coalition support, for the long haul.

But the surge strategy was undermined from the outset, when President Obama—in the very speech in which he announced the campaign in December 2009—also announced that it would end in July 2011. I happened to be sitting with two pro-Taliban negotiators from Kandahar when the surge was announced; as the televised speech ended, they shook their heads in amazement. 'Are you deliberately trying to lose?' one asked incredulously. The other pointed out that 'tomorrow, the Taliban will go out and say to the people "You heard the American President's speech. You might like the foreigners, but they are leaving in July. What will you do in August, when the foreigners have gone and we are still here?"' (Kandahar Taliban representatives 2009, Cedars Restaurant, Kabul, 1 December). In the event, our departure was delayed to December 2014, but the damage was done: by setting a firm timeline for our drawdown and then sticking to it regardless of conditions on the ground, we very helpfully told the Taliban exactly how long they had to survive in order to wait us out.

Taliban commanders did indeed wait us out. They needed to maintain a certain level of effort inside the country in order to stay in the fight, but throughout the surge they carefully calibrated their activity and withheld their main effort, keeping their leadership and their best fighters safe in Pakistan and building them up in anticipation of our departure. In effect, they treated some parts of Afghanistan as an economy-of-force operation. As the surge drew to a close in 2014, having damaged Taliban networks in

DOI: 10.4324/9781003317487-11

Afghanistan but done next to nothing to disrupt the parts of the structure safe in Pakistan, we congratulated ourselves on our success and justified our pre-planned departure on the basis that Taliban activity had diminished. In truth the Taliban were simply waiting until we left before escalating the conflict once again.

Under the post-surge SFA posture, as ISAF wrapped up to be replaced with Resolute Support from early 2015 onward, the Taliban did just that. ISAF combat brigades were replaced with Train, Advise and Assist (TAA) commands, and international forces stepped back from direct combat to a supporting role. Under the new approach, Afghan forces now carried the main combat burden and a smaller number of international troops were present primarily to train, advise and occasionally accompany Afghan police and soldiers as they sought to suppress the insurgency. Governance and development tapered off, the PRTs handed over to local actors, and—in what was probably a positive development, given its negative effect on corruption—aid spending dropped significantly. In addition to training and advice, under the SFA posture international forces provided air support as well as intelligence, surveillance and reconnaissance (ISR) capability to the Afghan forces, and international contractors delivered maintenance and logistic support, while coalition countries continued funding the effort.

After 2015, this approach was, in military and economic terms, quite sustainable for the United States and its allies. Coalition casualty rates fell dramatically from their peak during the surge, when fatalities averaged 34 per month, to just 1.5 per month from 2015 onward (iCasualties, undated). US spending on the war, likewise, dropped from a high point of $92 billion per year during the surge to about $58 billion per year. Hence, staying in Afghanistan would not have been unsustainable for the USA in economic or human terms, given the sufficient political will. But the fact that the coalition casualty rate dropped by a full 96% after the end of COIN, whereas the financial burn rate fell only 37%, highlights the harsh reality that we had transitioned from fighting the war until 2015 to mostly paying for it, with Afghans now doing the lion's share of fighting and dying.

And this was the critical problem: the post-2015 posture might have been sustainable for us, but it was far too costly in human terms to be sustainable for the Afghans. Total coalition deaths (including combat and non-combat fatalities, by all troop-contributing nations) during the entire war numbered 3,455, with the bulk of those occurring during the surge (iCasualties, undated). By contrast, Afghan police and military deaths were estimated at 66,000–69,000 over the same period, with most occurring after the surge, when Afghan forces bore the brunt of combat. This loss rate was simply too high to be sustainable for the Afghan forces over time—after 2015, the coalition could, in purely military terms, have kept the war going forever, but the Afghans could not.

An annual cycle emerged in the years after 2015, with a major Taliban offensive launched each spring seizing key territory, particularly in areas around the ring road, the highway network that connects Afghanistan's major cities with each other and Kabul. Each fighting season, the Taliban would push police and military garrisons out of isolated outposts in rural districts, and a government counter-offensive later in the year would recapture much (but crucially, not all) of that territory while simultaneously losing too many troops and police in the process.

Also, from 2015 onward—notably, during the Battle of Kunduz in September–October of that year—Taliban offensives tended to be larger, more sophisticated, and to target provincial capitals and other major cities. At the same time, asymmetric attacks (bombings, assassinations of government officials, drive-by shootings and raids on key urban sites) increased, forcing the Afghan military and police to divert significant resources to urban security. The appearance of Islamic State's Khorasan branch (ISIS-K)—a spinoff from the Pakistani Taliban that first arose across the south and east in 2015 and soon gained a solid foothold in the eastern province of Nangarhar on the Pakistani border—brought another, much more radical actor into the conflict, one that targeted (and was targeted by) the Taliban as well as the government, massively complicating the conflict.

In effect, with each successive year the Taliban grew stronger, foreign forces progressively pulled back, new combatants like ISIS-K emerged and Kabul's control in the countryside was eroding. The government was being forced to spread its troops ever more thinly to defend an increasing number of localities, and with each year's counteroffensive those troops were failing to recapture all the lost ground while suffering more casualties than they could afford. As a consequence, the government in Kabul found itself further and further behind the power curve at the start of each year's campaign.

The Selection-Destruction Cycle

One key reason for the unsustainable loss rate was, counterintuitively, the size and professionalism of Afghan Special Operations Forces (SOF). From 2011 onwards, as Afghan SOF grew, they became too big for the size of the overall Afghan defence establishment from which they were drawn. As the military sociologist Roger Beaumont pointed out in his 1974 classic, *Military Elites*, when an elite element such as a SOF component becomes too big in proportion to its parent organization, it can exert a net-negative effect on overall performance, through a phenomenon Beaumont called the 'selection-destruction cycle' (Beaumont 1974: 171–184). Capable troops, particularly NCOs, who would otherwise form the backbone of conventional combat units, are instead selected for service as individual fighters in elite units. Their skills, talent and motivation are thus lost to the larger force, which suffers a brain-drain from conventional units to its elite. Once in elite units, the force's best troops are thrown into high-intensity battles, used to rescue desperate situations or committed to dangerous tasks beyond the capabilities of ordinary troops, so that they typically suffer a significantly higher loss than their conventional counterparts. Over time, this selection-destruction cycle weeds out the most talented members from the larger force and kills them off, lowering the combat effectiveness of the organization as a whole.

From 2015, the selection-destruction cycle was running rampant in both the Afghan National Army (ANA) and the Afghan National Police (ANP). ANA SOF, and special ANP units such as the brigade commanded by Kandahar's PCOP, Abdul Raziq, were used as fire brigades—rushed around the country as quick-reaction forces to reinforce threatened areas, respond to Taliban or ISIS attacks, and spearhead the retaking of territory during each year's counteroffensive. The average ordinary ANA *kandak*, like the average ANP garrison, lost its best members to these high-priority forces, so ordinary units became less capable and less keen to fight. This in turn meant they required ever more frequent rescuing by SOF and other elite forces, further increasing SOF loss rates and accelerating the selection-destruction cycle. This dynamic was at its most intense during the final campaign of summer 2021.

A related issue was abusive behaviour toward the population by police and military units—driven by lack of resources at unit level, combined with high-level corruption within the ministries—which destroyed support for the government. Local ANP posts, for example, consisted in many areas of little more than one or two green-painted shipping containers tucked into the inside wall of a bulldozed earthen rampart surmounted by HESCO barriers, topped with razor wire, with a prefabricated cylindrical concrete guard-post on each corner. A beleaguered garrison might receive an occasional resupply of ammunition, fuel or radio batteries, but food (or even water in some parts of the country) were hard to come by, with resupply occurring intermittently at best. The pay was often late or diverted to corrupt senior officials so that garrisons lacked even the money to purchase supplies they needed, and frequently resorted to extortion from the community and robbing travellers passing through their checkpoints.

This abuse (and the contempt and disdain it generated from local people) undermined support for the army and police, made locals more willing to provide intelligence and support to the Taliban, and meant government forces in the districts were continually fighting to defend themselves, rather than rolling the Taliban back. Kabul's policy of not posting individuals to home regions—driven by fears of regional warlordism dating to the 1990s—further estranged the army and police from the people. And as SOF increasingly began rushing from crisis to crisis after 2015, the quick-reaction forces had even less connection with the population and became even more prone to alienate local communities, or engage in abuses themselves (HRW 2015).

The Afghan military and police inherited from ISAF a far more extensive footprint than they could sustain. They had too many small outposts in too many places, too few reserves, and were stretched to defend everywhere since the enemy could attack anywhere. But once an area had been occupied,

Kabul's prestige was committed and it was difficult to redeploy lest, by giving the appearance of abandoning a particular area, the government should lose legitimacy. This, indeed, was one reason why the government, encouraged by western advisers, insisted on recapturing as much lost terrain as possible at the end of each year's campaign season. But periodic abandonment and recapture of these contested areas—a pattern sometimes called 'mowing the grass'—had a devastating effect on local populations. Those willing to work with the military and police when they were present ended up being abandoned when they left. As the Taliban returned, these people had either to flee or be killed, often publicly and brutally, by the Taliban, as the insurgents sent a message to the rest of the local community. When the police and soldiers eventually came back, their previous allies would be dead, fled or intimidated, and the community would be even less willing than before to work with them.

In the last two years of the Obama administration, despite the steady erosion of government control in rural districts (and despite increasing violence and political instability in the cities) the coalition steadily drew down the numbers of advisers, contractors and aid personnel operating under the 'train, advise and equip' SFA mission. This reduction ran counter to best-practice COIN, in which withdrawal of combat troops and transition to SFA should be accompanied by an increase, not a reduction, in advisory support. It was driven by American domestic politics and was well underway by the end of the Obama administration. The incoming Trump team initially appeared open to supporting the Afghan government with a strategy of long-term commitment, a modest troop increase (to a total of around 15,000 troops in country) and an increase in air and drone strikes designed to intensify pressure on the Taliban to force them to negotiate (Nakamura 2017).

But by mid-2018 President Trump—never famous for his patience, and having campaigned on ending what he called 'ridiculous, endless wars'—was frustrated by lack of military progress on the ground, and reversed his policy, beginning direct talks with the Taliban over the head of the Afghan government, thereby conceding a key, long-standing Taliban demand (Chalfant 2017). By August 2019 he was calling for a drawdown as quickly as possible; by October that year the drawdown was underway. In February 2020 negotiators in Doha, led by Ambassador Zalmay Khalilzad, reached an agreement with the Taliban for a full withdrawal, and by the end of 2020 there were only 2,500 US troops in Afghanistan, down from 15,000 in June 2018, the month when Trump seems to have given up on trying to win the war and decided instead to withdraw (CRS 2021).

In this, despite initially seeming serious about actually winning in Afghanistan, Trump was consistent with previous US presidents in that he persisted in conflating the strategy in Afghanistan with the authorised 'troop ceiling' in theatre. Indeed, at times under every president since George W Bush it appeared as if the entire goal of American strategy in Afghanistan—and hence the parameters within which allies like Australia were forced to operate—was merely to reduce troop numbers on a domestic political timeline, ideally without any embarrassing defeat for Kabul or (far worse from Washington's standpoint) the need to put troops back into Afghanistan after pulling them out. It was this thinking that lay behind the Trump administration's desperate desire for a negotiated settlement with the Taliban, with or without the Afghan government, and led ultimately led to the February 2000 Doha Agreement, which side-lined the Kabul government and destroyed any leverage Afghan President Ashraf Ghani might have had, rendering him impotent.

As all this was happening, the Taliban's strategy of steadily choking the ring road, driving local police and military garrisons out and gaining control of rural districts through a combination of co-option, intimidation and delivery of local services (particularly mediation and dispute resolution services, and basic civil law and governance functions) was paying off. By early 2021, their approach of patiently surrounding district centres through successive spring offensives had put the insurgents into pole position for that year's campaign. When, in mid-April, President Joe Biden said it was 'time to end America's longest war' and announced that all U.S. troops would be out of Afghanistan by the twentieth anniversary of September 11th, Taliban leaders realised that the spring offensive they had already quietly launched might make this the final fighting season of the war (Smith 2021).

The Collapse of 2021

The campaign of 2021 began in early March with a series of small- and medium-sized attacks at the district level. Taliban commanders focused on picking off garrisons and district centres that were already isolated and under pressure. Local guerrillas, supported by higher-tier 'main force' Taliban units, partially encircled police and military posts, carefully leaving open an escape route for the garrisons. They then sent in a local elder, often a respected community member known to the garrison, to negotiate a surrender. Garrisons that agreed to hand over their posts along with their hardware and supplies were allowed to leave. The few that chose to fight were massacred to teach others an object lesson. Combined with torture and execution of any SOF or members of NDS (the Afghan intelligence service) unfortunate enough to fall into their hands, this meant that garrisons increasingly chose to flee rather than fight (Coren 2021).

Even for otherwise motivated commanders, the US decision to dramatically cut back air support, ISR, logistics and maintenance after May 2021 meant that local garrisons rapidly began to run out of ammunition, medical supplies, fuel, food and water. Helicopter resupply became increasingly impractical, while road convoys became fewer and smaller. Garrison commanders now knew they were on their own, while individual soldiers and police officers knew that if wounded there was no way they could be evacuated. The blow to morale was massively deepened when US forces abandoned their major airbase at Bagram in early July, without coordinating with Afghan counterparts, leaving outlying garrisons even more obviously out in the cold.

As increasing numbers of district centres fell during May, June and July, the Taliban began to encroach on provincial capitals and other urban centres, applying the same technique—partial encirclement, intimidation and negotiation to force a surrender. They also applied cash payments liberally as a way of convincing garrison commanders, mayors or even governors to stop fighting or switch sides. In this way, they succeeded in capturing ground and military equipment with little fighting and at the cost of few casualties. Psychologically, a narrative of unstoppable Taliban momentum began to take hold, which became self-reinforcing and ultimately self-fulfilling. There was also a network effect: for the few garrison commanders who still wanted to fight, it soon became obvious that with so many others surrendering, insisting on defending their bases would have made no difference to the outcome and simply resulted in their own people being brutally killed for nothing.

By mid-July, Afghan Commandos and other SOF units were fighting frantically to hold the key cities of Kandahar and Lashkar Gah, capital of Helmand. Here the selection-destruction cycle finally peaked, as Afghanistan's best troops—who could otherwise have stiffened every other unit in the ANA, formed a mobile reserve to enable others to hold the line, or consolidated to cover key cities including Kabul—were bled white in a desperate defence of the two southern centres. Simultaneously, with the ANA's best units concentrated in Helmand and Kandahar, Taliban columns began pushing on provincial capitals across the rest of Afghanistan. They attacked across the north, aiming to prevent the government from consolidating a stronghold in the northern third of the country, heartland of the ethnic and religious minorities (Tajiks, Uzbeks and Hazaras) who held off the Taliban through the Northern Alliance in the 1990s.

In the north, the policy of pushing assistance to the most dangerous and conflict-affected parts of Afghanistan had long infuriated local leaders and their communities. Pashtuns fighting the government in the south and east received billions in assistance, whereas communities in the north—which had supported the Northern Alliance against the Taliban in the 1990s, fought alongside US forces in 2001 and continued to oppose the insurgents through the first decade of the war—felt taken for granted. By 2015 much of the international assistance had dried up, leaving loyal populations in the north bypassed and embittered. This opened opportunities for a 'Northern Taliban' drawing support from detribalised Pashtuns and non-Pashtun ethnic groups, and by 2021 there was a strong Taliban presence, with popular support and local networks, in the region. Unsurprisingly, when the Taliban moved against government-controlled cities in the north in the summer campaign of 2021, locals felt little loyalty or commitment to defend the regime in Kabul or its northern outposts. All this set the scene for the final meltdown, the cascading collapse of one provincial capital after another, in August 2021.

After two decades, the final collapse took just nine days as the war appeared suddenly to speed up. First to fall was Zaranj, capital of Nimruz province in the remote southwest, which fell to the Taliban without much of a fight on 6th August; the next day Sheberghan in the north was captured after heavy fighting. Commandos, police and SOF in Kandahar and Lashkar Gah were now surrounded, under severe pressure, and US drones and carrier-based aircraft flew a few airstrikes to support them. These strikes, constrained by the need to fly enormous distances from Qatar or from aircraft carriers offshore due to the abandonment of Bagram the previous month, were largely ineffective and were some of the last coalition airstrikes of the war. Having lost its logistics and maintenance support, the Afghan Air Force at this point was almost out of rockets and bombs, less than two-thirds of its aircraft were airworthy and many pilots—intimidated by a carefully targeted Taliban assassination campaign against them and their families—were absent (Smith 2021, Stewart 2021).

Here, our tendency to conflate strategy with troop numbers came home to roost. Even accepting the impossibility (for both military and domestic US political reasons) of putting western troops back into Afghanistan to stabilise the situation, it would have been perfectly possible to surge air support as a way of bolstering the ANA. To be sure, the abandonment of Bagram made this dramatically more difficult, since aircraft had to fly enormous distances—from al-Udeid airbase in Qatar or from the decks of carriers in the Arabian Sea—to reach the battle area in Afghanistan, giving them very little time over their targets. It might also not have worked, of course. But in both material and morale terms, a surge in air support for the Afghan forces might have made a massive difference. Materially, coalition airstrikes would have allowed Afghan air forces to catch their breath, while letting ANA ground forces take advantage of some of the juiciest Taliban targets to appear on the battlefield at any time since 2001, as the Taliban main force massed in the open, in daylight, to attack the cities. In terms of morale, the voices of American pilots over the radio, speaking with ANA air controllers as they talked them onto those targets, would have sent a hugely important message of sustained support. Even a modest number of US advisers, including Special Forces Joint Terminal Air Controllers to control and direct air support, would have had an enormous effect on the outcome.

None of that happened, of course. The final US SOF teams to rotate into Afghanistan, drawn from 10th Special Forces Group (Airborne) based at Fort Carson, Colorado, left the country in the third week of June. Despite public protestations of unwavering support from President Biden, Secretary of State Tony Blinken and other officials—given until just two days before the fall of Kabul—from early August onward the coalition barely lifted a finger to help the Afghans stem the tide of Taliban victories.

Those victories now began to come thick and fast. Kunduz and Sar-e Pol, two northern provincial capitals, fell during the day on 8th August, followed late that evening by Taloqan, capital of the northern province of Takhar. The collapse of so many northern cities so quickly made it clear than any attempt to rally in a northern bastion—as had happened the 1990s—was certain to fail, even as Afghanistan's most capable forces, the SOF and Commandos, remained pinned down in the south. The next day, 9th August, the Taliban captured Aybak, capital of Samangan province, and the first major defection to the Taliban of a regional commander (Asif Azimi, a former Afghan senator) took place.

On 10th August two more provincial capitals—Farah in the west and Pol-e Chumri in the north—fell to the Taliban, and the next day Feyzabad, capital of the crucial northern province of Badakhshan, was captured. On 12th August, the city of Ghazni fell, cutting Kandahar off from Kabul by road, and the same day the major western city of Herat fell without a fight and Ismail Khan, one of the great anti-Taliban warlords of the 1990s, a former minister in the Kabul government and governor of Herat province, was filmed surrendering to the Taliban, in another stunning blow to morale. The same day Qala-e Naw, capital of the north-western province of Badghis, was captured.

Overnight, around midnight of 12th/13th August, Kandahar and Lashkar Gah both finally fell to the Taliban, with defeated Afghan commando units fleeing by air to Kabul from Kandahar (for a graphic depiction of the fall of Kandahar, see Vice News 2021). At this point the Taliban had captured so much military equipment, released so many prisoners and accepted the surrender of so many thousands of government troops that the government's ability to hold out was rapidly evaporating. US air support had dropped off to almost nothing, and having lost many of their best remaining troops in the defence of Lashkar Gah and Kandahar, the Ghani government was desperately short of forces to defend Kabul.

From Friday 13th August, the collapse accelerated even further. The fall of Kandahar and Lashkar Gah overnight was followed later that day with the loss of Chagcharan, capital of Ghor province, then Pol-e Alam, Tarin Kot and Qalat, capitals of the key provinces of Logar, Uruzgan and Zabul. After losing five provincial capitals on Friday, the government lost eight on Saturday: Asadabad, Gardez, Maimana, Mehtar Lam, Nili, Sharana, the critical northern city of Mazar-e Sharif and, late that evening, Maidan Shah. The only functioning forces on the government side at this point were a few isolated units outside Kabul, along with remnants of Afghan SOF and two understrength army divisions in the Kabul area. The final day of the campaign, Sunday 15th August, saw Jalalabad, capital of the eastern province of Nangarhar, fall early in the day, followed by Khost, Bamiyan, Mahmud-i Raqi, Charikar and Parun; Taliban troops took control of the airfields at Bagram and Kandahar the same day, and that afternoon they entered Kabul. In the end, the capital collapsed without a fight as Ashraf Ghani fled in disgrace, his government collapsed and the Afghan police and military melted away.

Why The Afghan Military Collapsed

As this account makes clear, the collapse of the Afghan government proceeded slowly and barely noticeably at first, beginning in 2015, but reached a tipping point somewhere in the spring of 2021. After that point, the war appeared to rapidly accelerate, leading to the catastrophic disintegration of the Afghan military over the summer, then the astonishingly rapid collapse of the Afghan Republic over a period of only nine days between 6th and 15th August. This, in fact, should not surprise anyone: it is entirely consistent with what we know of the endgame dynamics of internal conflicts. As a team led by Gordon McCormick at the Naval Postgraduate School demonstrated more than fifteen years ago, resistance movements degrade slowly but states—which embody many moving parts that all have to work together for government to function—collapse at an accelerating rate. McCormick's team observed that 'states generally pass a tipping point and enter their endgames and begin to decay at an accelerating rate. This is often an indicator that the final period of the struggle has begun. Between the time the conflict enters this phase and the time the state disintegrates, the conflict "speeds up" ... Insurgencies, by contrast, tend to decline historically at a decelerating rate' (McCormick 2007: 326).

Arguably, in a material sense, the withdrawal of US assistance in early May 2021 represented the tipping point of the campaign. The removal of air support, ISR, maintenance, advisers and contractors massively undercut the ability of the Afghans to perform resupply, maintain their vehicles and aircraft, evacuate their wounded or determine the enemy's next move, and—coming on top of years of unsustainable losses—the sudden cut in capability rendered the Afghan military effort unsustainable. It is worth noting in this context that, by pulling out these critical capabilities, the US and coalition governments did exactly what we had spent more than a decade assuring the Afghans we would absolutely never do.

The ANA, ANP and Afghan SOF were built like a tower of Jenga blocks: moderately stable while all the pieces remained in place, but completely unable to stand once certain critical capabilities were removed. Those critical capabilities—deliberately and by design—were provided by the coalition. Just to pick one example, in 2017–2018 when the aging Afghan helicopter fleet needed replacement the coalition, led by the US, insisted that rather than replacing existing Mi-17 helicopters with newer models of the same aircraft (which Afghan pilots already knew how to fly, and Afghan maintainers could service), the Afghans must instead adopt the much more complicated and sophisticated UH-60 Blackhawk. The new aircraft were far more complex to fly and maintain, and many of them turned up fitted for, but not with, the critical componentry needed to operate effectively in Afghanistan. But this was not the point: the goal was not to give the Afghans an aircraft they could independently fly and maintain, but rather to lock them into a US-centric, coalition-dependent platform that would prevent them acting independently (see for example Trevithick 2018 and Axe 2019). A similar dynamic operated across other capabilities: ongoing dependence on the international community (specifically, the Western coalition) for everything from vehicles and radios to weapons and ammunition was a feature, not a bug. Of course, the only reason Afghans were willing to agree to such an arrangement was that the coalition, led by the United States, promised repeatedly never under any circumstances to take these capabilities away. And of course, once they were whisked away regardless, following President Biden's withdrawal announcement, collapse was inevitable.

Even more importantly, the sudden removal of these critical capabilities had a ruinous moral effect on an ANA that already felt itself being betrayed by corrupt leaders in Kabul, whose soldiers could see politicians talking peace with the Taliban and organising their own escape from Afghanistan, while simultaneously pushing the troops into combat. When the coalition capabilities that underpinned the Afghan military's ability to actually fight were pulled out in May, the impact on morale was devastating. Then the Americans' departure from the critical Bagram airbase in early July—fleeing in the middle of the night, without so much as a courtesy call to Afghan counterparts, who found their 'allies' gone, the base abandoned and the electrical power cut off when they awoke in the morning—cemented the collapse of the air support capability on which the ANA depended, and massively reinforced the blow to morale.

Once several district garrisons had surrendered, the network effect took hold and more districts fell; once enough districts had fallen the Taliban moved on to target the provincial capitals; the narrative of inevitability began to sweep the country, and predictions of doom became a self-fulfilling prophecy. US promises of air support—given by Biden and Secretary Blinken as late as two days before the fall of Kabul—never materialized, while the pathetic call from American negotiators in Doha for the Taliban to spare the US embassy in Kabul (which became public on August 12th) emphasized that the United States had no plans to fight for its partner, but was now solely looking out for itself (Jakes 2021). Indeed, the very fact that these US negotiators, led by Zalmay Khalilzad, were engaged in talks with the Taliban on putting together a transitional government in Doha even as Ashraf Ghani's government was fighting for its life in Kabul showed that Afghanistan's so-called allies were more than willing to allow a Taliban takeover. The message to the fighting men on the front lines around Kabul could not have been clearer.

In the final hours, however, it was the collapse of the civil government—and specifically the fleeing of Ashraf Ghani and his inner circle from the Arg in central Kabul—that convinced those combat units still operating in and around Kabul that all was lost. It's therefore worth asking why the civil government collapsed so abruptly.

Why The Civil Government Collapsed

In addition to all of the factors mentioned so far, the isolation of President Ghani himself is worth mentioning. The president, who had failed to garner more than a small percentage of the vote in either of his previous elections, achieved an even smaller percentage in one of the lowest turnout elections in Afghan history, in 2019. Ghani had minimal support from the population, his support base lying largely in the urban elite and the international community. Yet he was also strangely disconnected from the rest of his government, his irascibility and pedantic nature bringing high turnover at senior levels of the Afghan administration, as officials quit or were fired at astonishing rates. Likewise, provincial and local officials in many cases had minimal ties to their regions; drawn from a rentier elite that relied on international support for access to the resources that protected their power, they had nothing to offer their communities when that support was withdrawn.

Almost to the end, Ghani saw regional commanders—Ismail Khan, Rashid Dosum, Noor Mohammed Atta—and their local militias as a threat to the central government's monopoly of force, and so side-lined and starved them of support until, too late, in the final weeks of the war, he realised his government could not both defend its territory and generate a mobile reserve with sufficient strength to confront the Taliban. Two weeks before the fall of Kabul, the warlords finally achieved a measure of official status, their militia were re-designated as Public Forces, and a joint command centre was created to integrate their efforts with those of the ANA and ANP. It was much too late: ill-equipped and untrained, the militias were thrown into battle during the final days of the war, unprepared and unsupported. Most melted away within hours, and commanders like Dostum and Atta were quickly forced to flee across the border into Uzbekistan.

Crisis reveals character. Ghani's was focused on the administrative trappings of government and the intellectual, not to say bookish, pursuits of the international class from which he came, rather than on the substance of government presence, personal contact with the Afghan people and the delivery of a social contract with the population and with regional and local communities. Once districts and provinces began to fall, Ghani's aloofness left him in an isolated bubble. To those who met with him in the last weeks of the war he appeared frozen, in shock or in denial, interacting only with a tiny circle of close advisers. But there was an undertow of panic, just below the surface, which broke out when it was believed that the Taliban had entered the palace. Ghani fled with some—though not all—of his advisers, his government collapsed immediately, and the military (the hard-core remnant of which had been preparing to defend Kabul until that moment) found itself with nothing left to defend, and melted away. The Taliban forward troops had paused on the edge of the city, hoping for an orderly transition of power in which they would not need to occupy Kabul until the Western evacuation was complete, but the rapid collapse of Ghani's government prompted US leaders to ask them to come into the city and occupy it. The deadly chaos around Hamid Karzai International Airport—soon, one assumes, to be renamed—was the result.

What Could We have Done Differently?

There are literally a million things the international community, and the US-led coalition forces, could have done differently had we cared enough to avoid the catastrophic endgame of 2021. For reasons of space, I'm not going to go into these in detail. Suffice it to say that missed opportunities to make peace earlier in the campaign when we were still in a strong position were a key part of our defeat. Likewise, we could have built an Afghan military that mirrored its parent society and reflected how Afghans (some of the world's most accomplished fighters) prefer to fight, rather than mimicking our own way of war. Instead of adopting the 'red first' aid strategy—where we poured aid into the most violent and unstable places in Afghanistan in hopes of stabilising the environment—we could have adopted a 'green first' strategy of building on success. A focus on local government, applying resources to the district level in support of provincial and rural development, would have been another thing we could have done differently. We could have put the rule of law at the centre of our efforts, rather than treating it as an afterthought or supporting effort.

Obviously, we could have kept our promises to the Afghans that we would maintain our support for them after the end of the COIN period and the transition after 2015 to SFA. Having designed specific dependencies into the Afghan system, somehow senior leaders and politicians seem to have forgotten their own earlier decisions and expected a system to stand on its own two feet that we had specifically designed not to be able to do so. Reducing troop numbers could have worked, if it had been accompanied with a surge of materiel, logistics, air strikes, intelligence support and (perhaps) a regional reserve poised to deploy in extremis as required. None of that happened: instead, we drew down every aspect of support simultaneously and were then shocked, shocked to discover that the Afghan military we had designed not to be able to fight without our help could not do so.

At the political level (and relevant here because it was the collapse of Ghani's government that ultimately triggered the disintegration of the Afghan military) we could have insisted on fair elections and not interfered to pick the winners we wanted to see rather than those that Afghan voters preferred. This would have avoided the manipulated stalemates, broken only by international pressure, that occurred in 2009 and 2014, or the chaotic mess of 2019 in which two rival candidates claimed victory and held competing inauguration ceremonies. The legitimacy and credibility of the Kabul government—in the eyes of its own people, not its international donors—was the critical missing piece for much of the war. The Taliban, who were nothing if not authentic, had it all over us as a result. Had we possessed a legitimate partner with popular support in the Afghan government, we could have negotiated differently. Rather than doing so in such a way as to sideline and render impotent the Afghan government, negotiating over their heads for a separate peace that left them hanging, we could have structured our negotiating approach to force the Taliban to deal with Kabul on the Afghan government's terms. By making the Kabul government so dependent on ourselves, and then wanting to leave more than we wanted them to win, we doomed them to destruction and ourselves to defeat.

Finally, and by far the most important, we could have stopped persistently underestimating the Taliban. Perhaps it was the swiftness of our victory in 2001, or their lack of material capability for most of the war, or their traditional garb and rustic language, but whatever the reason—across two decades, and every aspect of the effort in Afghanistan—we consistently underestimated the Taliban. As a fighting force, a political organisation, a nationalist propaganda critique of the government, a parallel governance structure or a social movement, we never took the Taliban seriously at any time during the war—not, at least, until they took our partner government down and captured Kabul.

As late as July 8th 2001, five weeks before the fall of Kabul, President Biden was scoffing at the notion that the Taliban could defeat the government. Ashraf Ghani, for his part, spent his last weeks in office focused on e-governance, the fourth industrial revolution and public service reform—this while the Taliban were killing hundreds of his troops, overrunning dozens of garrisons and capturing multiple districts per week. A series of military commanders and civilian officials were still obsessing over secondary, nice-to-have issues when the enemy was already at the gates. Taking the Taliban seriously, as the existential threat they became, would have made the most important difference of all. Unfortunately, we never did, and the endgame of 2021 was the result.

Conclusion

In the final analysis, the collapse of the Afghan military principally resulted from lack of will. Neither Afghans nor Americans went down gamely, fighting to the last: rather, the campaign of 2021 sat somewhere between a surrender and a betrayal. It was one of the most momentous foreign policy disasters since the 1970s. Indeed, in many ways, it combined the US defeat in Saigon in 1975 with the Iran hostage crisis of 1979-81, but on a far grander scale: the thousands of American citizens, Green Card holders and Afghan allies left behind are effectively now hostages at the mercy of the Taliban.

Our defeat in Afghanistan was a massive morale boost for every jihadist group on the planet. After two decades, the scrappy little Taliban guerrilla movement had succeeded in defeating the all-powerful United States through persistence, continued belief in the one true God, and maintaining faith in the ultimate victory of their cause. Within the country itself, however, something one hears both from pro-Taliban and even from former government supporters is to note, with a sense of wry pride, that Afghan insurgents have now defeated not one but two superpowers in a single generation: the Soviets in 1989 and the United States in 2021. As I have tried to show here, the endgame dynamics of a conflict like Afghanistan are thus well worth studying.

Note

1 Portions of this chapter appear in David Kilcullen and Greg Mills, *The Ledger: Accounting for Failure in Afghanistan* London, Hurst & Co., 2021.

References

Axe, D., 2019. 'The Pentagon Forced Afghanistan to Take Inferior Helicopters' *The National Interest*, 5 February https://nationalinterest.org/blog/buzz/pentagon-forced-afghanistan-take-inferior-helicopters-43297

Beaumont, R.A., 1974. *Military Elites.* Bobbs-Merrill Company, Indianapolis/New York.

Chalfant, M., 2017. 'Trump knocks "ridiculous Endless Wars" amid US troop pullout from Syria' *The Hill*, 19 October https://thehill.com/homenews/administration/464612-trump-knocks-ridiculous-endless-wars-amid-us-troop-pullout-from-syria

Coren, A., Sidhu, S., Lister T. and Bina, A.B., 2021. 'Taliban fighters execute 22 Afghan commandos as they try to surrender,' *CNN*, 14 July https://edition.cnn.com/2021/07/13/asia/afghanistan-taliban-commandos-killed-intl-hnk/index.html

CRS, 2021. *U.S. Military Drawdown in Afghanistan: Frequently Asked Questions*, Congressional Research Service, 4 February https://crsreports.congress.gov/product/pdf/R/R46670

HRW, 2015. *'Today We Shall All Die': Afghanistan's Strongmen and the Legacy of Impunity*. Human Rights Watch. 3 March https://www.hrw.org/sites/default/files/report_pdf/afghanistan0315_4up.pdf

iCasualties, undated. Afghanistan casualties per year, iCasualties Iraq: iCasualties Home Page.

Jakes, L., 2021. 'U.S. Asks Taliban to Spare Its Embassy in Coming Fight for Kabul' *The New York Times*, 12 August https://www.nytimes.com/2021/08/12/us/politics/taliban-afghanistan-us-embassy.html

McCormick, G.H., Horton S.B. and Harrison, L.A., 2007. 'Things Fall Apart: The Endgame Dynamics of Internal Wars,' in *Third World Quarterly*, vol. 28, p. 2.

Nakamura, D. and Phillip, A., 2017. 'Trump announces new strategy for Afghanistan that calls for a troop increase' *The Washington Post*, 21 August https://www.washingtonpost.com/politics/trump-expected-to-announce-small-troop-increase-in-afghanistan-in-prime-time-address/2017/08/21/eb3a513e-868a-11e7-a94f-3139abce39f5_story.html

Rahi, A., 2021. 'Afghan Forces are Not Losing, They are Disintegrating' *The National Interest*, 1 August https://nationalinterest.org/blog/buzz/afghan-forces-are-not-losing-they-are-disintegrating-190730

Smith, S. and Egan, L., 2021. '"It is time to end America's longest war": Biden announces full withdrawal of troops from Afghanistan,' *NBC News*, 14th April https://www.nbcnews.com/politics/white-house/it-time-end-america-s-longest-war-biden-announce-full-n1264048

Stewart, P., Ali, I. and Shalizi, H., 2021. 'Special Report: Afghan pilots assassinated by Taliban as U.S. withdraws,' *Reuters*, 9 July https://www.reuters.com/world/asia-pacific/afghan-pilots-assassinated-by-taliban-us-withdraws-2021-07-09/

Trevithick, J., 2018. 'Pentagon Admits Afghanistan's New Black Hawks Can't Match Its Older Russian Choppers' *The Drive*, 15 June https://www.thedrive.com/the-war-zone/21558/pentagon-admits-afghanistans-new-black-hawks-cant-match-its-older-russian-choppers

Vice News, 2021. *The Fall of Kandahar*, 24th August https://youtu.be/vGQ34RCv7Pc

9 Some Reflections on The Pursuit of 'Peace' in Afghanistan: 'Never Send to Know for Whom the Wars End'

William Maley

Afghanistan is an unusually complex social space. Even before the communist coup of April 1978 that triggered decades of conflict, scholars had long marked the diversity of the Afghan population. Overwhelmingly Muslim, it nonetheless comprised more than 50 different ethnic groups, and also reflected the schism found elsewhere in Muslim countries between Sunni and Shiite Muslims. The bulk of the population lived in rural areas and engaged in agricultural or pastoral activity, or commerce in support of primary industry. The communist coup and the Soviet invasion of Afghanistan in December 1979 brought major changes. Out of a pre-war population estimated at 13.05 million, more than 6 million were displaced as refugees into neighbouring countries, predominantly Pakistan, with dramatic implications for human capital formation and for the maintenance of infrastructure within Afghanistan itself (Ibrahimi 2020: 3). Having encountered fierce resistance from so-called *Mujahideen* groups, a new Soviet leadership under Mikhail Gorbachev withdrew Soviet combat forces by February 1989, and the Communist regime finally collapsed in April 1992. But what followed was more disarray, as Pakistan, long a host to refugees and *Mujahideen* parties, sought to promote its own clients to positions of power. In September 1996, Pakistan's last proxy, the anti-modernist Taliban movement, was able to seize the Afghan capital Kabul from its exhausted opponents, but the Taliban regime lasted only until it was overthrown by the US and its allies following the September 2001 terrorist attacks in the United States (see Maley 2021a: 202–215). It took the Taliban nearly twenty years to reacquire power. How this came to pass is the subject of this chapter.

Much of the analysis offered in the following discussion focuses on very recent developments, and particularly the consequences of almost unbelievably maladroit diplomatic activity on the part of successive US administrations. But that said, there were also significant problems contributing markedly to a long-term decline in stability within Afghanistan, problems that dated right back to the aftermath of the Taliban's initial overthrow in 2001 (Jamal 2022). One of the most serious was the adoption in the 2004 Afghan constitution of a strong-presidential model of government. This was anything

DOI: 10.4324/9781003317487-12

but a brilliant idea in a country with so many ethnic groups, not one of them comprising an absolute majority of the population since it gave rise to the likelihood that one group would see itself as a big winner and all others would see themselves as losers. Furthermore, while there were plenty of candidates for public office who had *some* of the skills that one might wish to see in a strong president, there were virtually none who had *all* the necessary skills. This set the scene for mounting disappointment about the performance of incumbent leaders. Finally, the presidency was severely overloaded, with the occupant expected to be the symbolic head of state, executive head of government, and a one-person interagency process to reconcile tensions between different components of the state. Each was potentially a full-time position; combining them in a single office was a recipe for underperformance.

Added to this was the problem of a highly centralised state, but one that was administratively fragmented. The Bonn Agreement of December 2001, signed by non-Taliban Afghan political actors, provided for up to 29 'departments' in an interim administration when probably 6 to 8 were the most that were required. This set the scene for interagency rivalry, as different political groups tended to control different departments. Under the new 2004 Constitution, district and provincial governments were notably weak, essentially branches of the central state rather than sources of localised legitimacy. This, over time, created a sense of disconnect between rulers and the ruled. Moreover, especially during the rule of President Hamed Karzai (2001–2014), the political system increasingly took on a 'neo-patrimonial' character (Maley 2018: 37–40), with formal institutions entwined with patronage networks from which some benefited enormously, but from which others were excluded. The salience of networks (Sharan 2023) fuelled corruption, although the reckless supply of aid monies, far exceeding the absorptive capacity of the new Afghan state, also fostered this development.

In addition, external factors complicated Afghanistan's transition. The United States, nominally Afghanistan's major supporter, was increasingly distracted from 2003 by its invasion of Iraq. This was to prove immensely damaging for Afghanistan; the invasion of Iraq fundamentally compromised the momentum of Afghanistan's transition. With the situation deteriorating, the new Obama administration in 2009 embarked on a 'surge' in Afghanistan that was designed to break the momentum of the Taliban's own resurgence which Pakistan had supported. It proved, however, of limited value, since President Obama at the *outset* of the surge also identified the point in time at which it would begin to be wound back; it was thus a time-based rather than conditions-based exercise, and simply invited the Taliban to regroup once the US troop withdrawal began. Afghanistan survived the subsequent scale-down of the US's military presence, which by the end of 2014 was relatively small, although a source of critical niche capabilities. This was partly because the scaling-down was widely presented as a cooperative undertaking of both the US *and* the Afghan government. The problem of insurgency, however, had not gone away.

This had much to do with the geopolitics of the region. Pakistan, desperate to prevent the consolidation within Afghanistan of a government that could be pro-*Indian*, quickly resumed support for the Taliban as a proxy for its own interests, providing sanctuaries, equipment, and logistical support, all crucial to the sustaining of the Taliban's attacks on targets in Afghanistan and designed to weaken the Afghan government by symbolising its inability to provide protection for those under attack. At the same time, the character of the Taliban as a terrorist organisation was fully on display: if a terrorist organisation is one that uses actual violence against non-combatants for a political purpose with a view to creating a psychological impact that is disproportionate to the physical harm caused, then the Taliban behaved in such a way as to tick every box (Maley 2021b). But Western governments, most notably that of the United States, were reluctant to characterise the Taliban in this way, not least because the idea that it could be fruitful to negotiate with them at some point had long enjoyed traction in some Western circles. This came to a head under the administration of US President Donald Trump.

The Doha Agreement and its Defects

On 29 February 2020 in the city of Doha in Qatar, a US envoy, Dr Zalmay Khalilzad, signed an 'Agreement for Bringing Peace to Afghanistan' (*Mowafeqatnamah-e awardan-e solh ba Afghanistan*) with the Taliban. The Afghan Government was not a party to the Agreement, and had not been a participant in the negotiations that led to its conclusion. It contained no provision for a ceasefire, no provisions protecting women's rights, minority rights, or human rights more generally, and no commitment to Afghanistan's constitutional framework. It did, however, contain a timetable for the total withdrawal of all forces of the United States, 'including all non-diplomatic civilian personnel, private security contractors, trainers, advisors, and supporting services personnel', as well as the forces of its 'allies and the Coalition'. Furthermore, it provided (as a 'confidence building measure') for the release by 10 March 2020 of 'up to five thousand' Taliban 'combat and political prisoners', who were actually in the custody of the Afghan Government rather than the US. The prisoner-exchange commitment that the US made to the Taliban had lethal effects in multiple spheres. It ensured that the US would be neither trusted by the Afghan Government nor feared by the Taliban and sent a terrible message to the wider Afghan public, which face the immediate threat of battle-hardened Taliban resuming combat activities, which was exactly what they did. It was very clear that the Agreement constituted an exit agreement for the United States rather than a peace agreement for Afghanistan. And that was precisely how things panned out (Maley and Jamal 2022).

Several features of the agreement marked it as a strikingly amateurish exercise. One was the problem of 'floating texts.' It is a commonplace proposition in diplomacy that if one wishes to avoid destructive ambiguity, it is important that the wording of key texts be carried out with the greatest of care, in order to avoid the sending of contradictory signals. On the same day that the Doha agreement was signed, a 'Joint Declaration' was issued in Kabul by the Afghan government and the United States. This was distinct from the Doha Agreement, and was not legally binding, but it did address some issues with which the Doha Agreement was concerned. Critically, however, it did so using different wording, exactly the kind of failure that professional diplomats strive to avoid. In discussing prisoner releases, the Joint Declaration simply stated that the 'Islamic Republic of Afghanistan will participate in a US-facilitated discussion with Taliban representatives on confidence-building measures, to include determining the feasibility of releasing significant numbers of prisoners on both sides.' As if this were not sufficiently confusing, there was also much talk about 'secret annexes' to the Doha agreement, but on 1 March 2020, US Secretary of State Pompeo described the previous day's text as 'the complete agreement', a remark which had the effect of liberating the Taliban from any commitments in unpublished materials.

Five severe defects marred the Doha Agreement and went a long way to explaining its failure to bring peace to Afghanistan. First, it was fundamentally premised on the defective assumption that the Taliban were interested in engaging in good-faith negotiation. As Kate Clark of the Afghanistan Analysts Network observed, 'Khalilzad had gambled all on the Taleban genuinely wanting to negotiate. He never had a Plan B of what to do if the insurgents were playing for time and were actually intent on military conquest. Bizarrely, he and other US officials clung to their fantasy peace process into August, even as huge swathes of Afghanistan fell to the Taleban' (Clark 2021b). Second, the US approach was based on a defective process: initially it adopted the perfectly defensible position that 'nothing is agreed until everything is agreed,' but it abandoned this in favour of a two-stage process in which the US would first strike its own agreement with the Taliban. The risk of this, which in fact materialised, was that if the Taliban felt that they had achieved through the agreement with the US everything that they really wanted, they would have no incentive to negotiate seriously with the Afghan government. The result was that the much-vaunted 'intra-Afghan negotiations' proved worthless, with the Taliban simply engaging in strategic stalling. Third, the US engaged in defective signalling. Its desperation to exit Afghanistan invited the Taliban to increase the demands during the 'implementation phase' of the agreement, something which became very apparent when they insisted on the release of prisoners who were criminals (such as the rogue Afghan Sergeant Hekmatullah, who murdered three Australians who were *hors de combat* in their compound) rather than 'combat' or 'political' prisoners. The US, which buckled in the face of this pressure, had clearly forgotten the warning that if one acts as if one can be taken for granted, one *will* be taken for granted. Fourth, the US displayed a culpably defective understanding of mass psychology in Afghanistan. It does not pay to be on the losing side in Afghanistan. As Hobbes put it, 'Reputation of power, is power' (Hobbes 1996: 62), and the US-Taliban agreement boosted the reputation of the Taliban and seriously undermined the standing of the Afghan government: in a very real sense, the Doha Agreement lit the fuse for the collapse that occurred in August 2021. And finally, the Doha Agreement was premised on a radical misunderstanding of the conflict in Afghanistan, which US officials were inclined to frame simply as a conflict of *interests*, fatally overlooking the extent to which it was also a conflict of *values*, with the ideology of the anti-modernist Taliban fundamentally irreconcilable with the 'Republican' model that had taken shape after 2001. While some experienced analysts warned of this defect, other analysts, often less experienced than they thought themselves to be, focussed on building what Clark bluntly called 'fantasy castles of research, advocacy and new institutions' (Clark 2021a), with lamentable consequences.

An 'Unwinnable War'?

Why was the US so keen to exit Afghanistan? It had long since ceased to be an especially costly commitment for the United States. Between the end of 2014 and the collapse of the Republic, only 64 US military personnel were lost in combat situations in Afghanistan, and the cost of the war in 2020 was only 0.006% of US federal outlays. US domestic politics, reflected in the isolationisms of both President Trump and President Biden, seemed to have played a much larger role than any detailed calculations of cost. But beyond this, much of the impetus for withdrawal seemed to have sprung from the framing of the conflict as 'unwinnable.' This, however, depended on a particular notion of war as an activity to be terminated by the battlefield victory of one party and the defeat of another. Such a simple notion would have surprised Thomas Hobbes, who famously wrote that 'the nature of War, consisteth not in actuall fighting; but in the known disposition thereto, during all the time there is no assurance to the contrary' (Hobbes 1996: 88–89). Hobbes's approach implied that much more may be required than simply battlefield developments in order to terminate war, and that physical extraction of one's forces may not achieve that particular outcome, even for a party that is cutting and running.

Here, it is worth noting how historical images of war termination can shape leaders' understandings of what is possible or desirable. In the US, there are powerful images of this kind at play. The surrender of General Robert E Lee of the Confederacy to General Ulysses S Grant of the Union at Appomattox in 1865 has proved as canonical as the images of Japanese officials surrendering to General Douglas MacArthur on the USS Missouri in Tokyo Bay in 1945. What is striking about these examples, however, is that they involved either state-to-state conflict (as in the Second World War), or conflict between a state and a proto-state (as in the American Civil War). By no means all wars of this character. Some civil wars, for example, are much more ragged in their nature, and end by sputtering out rather than with the kind of formal ceremonies that occurred in 1865 and 1945. The Russian Civil War after the 1917 Bolshevik Revolution provides an obvious example. Indeed, in an era in which the 1945 *Charter of the United Nations* imposes a specific prohibition on the use of force except in self-defence or when authorised by the UN Security Council, the old-style model of general surrender is now the exception rather than the rule.

State-to-state warfare is not obsolete, as Ukraine's stalwart resistance to the Russian invasion of February 2022 made clear, but many other kinds of warfare have grown in significance since the end of the Second World War. Guerrilla warfare, witnessed in parts of Africa and Southeast Asia, has posed significant challenges to both state authority and the system of states. Regular armies struggled to cope with such enemies, with varying degrees of success, during the Malayan Emergency, the Vietnam war and many similar episodes of strife. The phenomenon of insurgency, linked in some analyses to 'the "spirit" of traditional peasant "rebellion"' (Desai 1990: 442), prompted the emergence of a substantial literature on 'counterinsurgency' which attracted a great deal of attention during the Obama-era 'surge' in Afghanistan, as observers debated the merits of various strands of counterinsurgency doctrine (see Kilcullen 2009; Kilcullen 2010). Mary Kaldor in particular drew attention to what she saw as the phenomenon of 'new wars', marked by distinctive actors, methods, goals, and resourcing (Kaldor 2012). Furthermore, in determining how to counter violence, policymakers increasingly confronted the challenge of moving from counterinsurgency to counterterrorism (Maley 2016), with the latter involving not obvious victories, but a shift to policing and community development as significant tools with which conflict could be de-escalated. In this kind of context, a 'Tokyo Bay' conception of war termination could appear quite antiquated.

In this kind of conflict, is the concept of 'winning' meaningful at all? On the one hand, a *narrative* of 'victory' can be a powerful tool in mobilising a population against an aggressive enemy. This was something that Winston Churchill grasped in 1940, in a way that his predecessor of British Prime Minister, Neville Chamberlain, totally failed to do: for Chamberlain, a positive outcome in the war was to flow from a realisation by Nazi Germany of its economic costs. Churchill, far more realistic in his appraisal of Hitler, understood that this would never be the case and his advocacy of victory at all costs was exactly what the British public wished to hear. But, of course, Churchill's advocacy was a reflection of the nature of the conflict in which Britain was actually involved. A very different kind of conflict might well have inspired—or required—a very different kind of rhetoric. In guerrilla warfare or insurgency, such 'command structures' as the enemy possesses may be quite obscure, as a result of which it may be unclear who (if anyone) even has the authority to offer 'strategic surrender' as opposed to the surrender of a few localised combatants. The complete obliteration of an enemy, even in the absence of a formal surrender, might be seen as a form of victory; but complete obliteration may also be very hard to achieve.

One other point is worth noting in passing: the US involvement in Afghanistan before the end of 2014 was very different from its involvement thereafter. Indeed, one might even raise the question of whether the United States was 'at war' in Afghanistan after 2014 in any strong sense of the term. The mere presence of American troops is hardly decisive of the question: large numbers of US troops have historically been deployed without being involved in active combat. Furthermore, the number of US 'war casualties' from 2015 to the 15 August 2021 collapse in Afghanistan, as noted earlier, was very small at just 64, and dwarfed by the number of suicides amongst 'Active Component' US military personnel over the same period, namely 2068 (see Orvis 2021: 4). To put it another way, *for the US* the war in Afghanistan largely ended in 2014: thereafter, it was prosecuted by Afghan forces, with US personnel mainly in non-combat support roles, or operating drones from remote locations. For the Afghans it was a different story: from 2015 onwards, it was they who bore the responsibility for the heavy lifting.

US Framing

How the leadership of a power *frames* a conflict can have a great deal to do, both militarily and diplomatically, with how it is prosecuted and terminated. Framing 'refers to the process by which people develop a particular conceptualization of an issue or reorient their thinking about an issue' (Chong 2007: 107). It is, of course, possible for different actors within a state to seek to 'frame' a conflict somewhat differently, but once a particular frame becomes dominant, these differences are likely to be subsumed into a wider strategic narrative. One significant problem associated with the US approach to Afghanistan was that civilian politicians in the US were keen to avoid framing the conflict in Afghanistan in a way that highlighted Pakistan's role in supporting the Taliban. Politicians were all too aware of this role, but at the same time readily swallowed the line that any pressure on Pakistan risked producing an internal collapse in that state.

The consequence was that the US framed the conflict in Afghanistan as an internal conflict rather than 'creeping invasion' or proxy war (Farasoo 2021) and, at least for a while, focussed on counterinsurgency rather than more holistic approaches in trying to address the challenges that the Afghan government faced from the Taliban. A creeping invasion occurs when a middle power uses force against the territorial integrity or political independence of another state, but covertly and through surrogates, denying all the while that it is doing any such thing; and this use of force is on a sufficient scale to imperil the exercise of state power, by the state under threat, on a significant part of its territory, and is designed and intended to do so (Maley 2002). When a state is targeted in a creeping invasion, it will rarely be sufficient to respond simply with military activities on that state's territory. Rather, an integrated military and diplomatic response will be required.

The importance of diplomacy in war can easily be overlooked. This is partly because in classic 'state-to-state' war situations, engagement with the enemy may be dormant until the moment of surrender, although managing relations between allies can be an extremely important undertaking. But in conflict such as that in Afghanistan, where Pakistan, nominally partnering with the United States, was actually pursuing a 'two-track' approach (Weinbaum 2008; Alexander 2021), diplomacy may be critical as a tool for addressing such perfidy. There are several reasons why this was the case with respect to Afghanistan. First, the problem was an absolutely critical one. In November 2009, the US Ambassador to Afghanistan, retired Lieutenant-General Karl Eikenberry, warned that

> More troops won't end the insurgency as long as Pakistan sanctuaries remain. Pakistan will remain the single greatest source of Afghan instability so long as the border sanctuaries remain, and Pakistan regards its strategic interests as best served by a weak neighbor ... Until this sanctuary problem is fully addressed, the gains from sending additional forces may be fleeting.
>
> (Eikenberry 2009)

Second, Afghanistan was in no position to invade Pakistan in order directly to attack the Taliban's sanctuaries. It lacked the military capacity to do so and US politicians would have lent no support to such activities. Third, the US military was not able to mount such attacks either. There is no general right of 'hot pursuit' on land, and there was no support from Washington to address the problem by military means. Fourth, it was Pakistan that had to be the target of pressure. Talking with the Taliban offered no solution to the problem when the existence of sanctuaries had the effect of insulating the Taliban from any need to negotiate seriously. The old question of 'why talk to the monkey when you can talk to the organ grinder' was particularly pertinent in this situation.

Unfortunately, political figures in Washington proved unequal to the challenge of responding to the key points that Ambassador Eikenberry had raised, and instead opted to frame Pakistan as part of the solution rather than as part of the problem. This, *inter alia*, highlighted the limitations of framing as a tool for dealing with unpleasant problems, and in a way reminded one of Richard Feynman's famous observation following the 1986 Challenger disaster that 'For a successful technology, reality must take precedence over public relations, for nature cannot be fooled' (Feynman 2001: 169). This problem became even more acute with the Obama-era 'surge' of activity in Afghanistan. With the problem of sanctuaries unaddressed, the 'gains' from the surge proved fleeting, just as Eikenberry had predicted, and the surge itself, ironically enough, heightened the dependence of the US on Pakistani cooperation since logistical support for the expanded troop presence in Afghanistan had to be supplied through Pakistani territory. In the absence of vigorous and orchestrated diplomatic pressure on Pakistan, making use of instruments that specialists had no problem in identifying (Haqqani 2017), the 'surge' was no recipe at all for dealing with the Taliban threat.

It should also be noted that the United States framed the *Taliban* in a way that reflected a fundamental misunderstanding of their character. This was most obviously displayed in October 2020, as Taliban attacks escalated across Afghanistan and Taliban officials ostensibly taking part in 'intra-Afghan talks' engaged in strategic stalling. On 19 October 2020, Dr Khalilzad, an active user of social media, 'tweeted' that 'Continued high levels of violence can threaten the peace process and the agreement and the core understanding that there is no military solution'. By this time, any notion that the Taliban shared such a 'core understanding' was manifestly delusional.

'Not Losing' as A Feasible Objective

Arguably one of the worst mistakes the United States made in Afghanistan was to lose sight of *not losing* as a useful strategic objective. At first glance this may seem quite peculiar: would any party embark on a war with such a modest objective? But on closer scrutiny, it makes rather a lot of sense to take 'not losing' seriously as a strategic outcome of choice. Few politicians would be likely to make 'not losing' a central element of their rhetoric, and as we have seen, Churchill opted for exactly the opposite in 1940. That should not, however, disguise the fact that between the fall of France in June 1940 and the Japanese attack on Pearl Harbor in December 1941, Churchill would have been extremely hard-pressed to trace any credible pathway for a victory against Nazi Germany on the part of the United Kingdom and its Commonwealth partners. At the very time that Churchill was publicly proclaiming victory as a strategic objective, his *actual* objective was to avoid losing. This was entirely legitimate, and vastly more desirable than the alternative that was promoted in late May 1940 by the Foreign Secretary, Viscount Halifax, of exploring options to negotiate with Berlin (see Lukacs 1999: 113). Churchill appreciated, as Halifax did not, that what was required to defeat Germany was the sort of parametric shift that occurred when the United States entered the war; and that in the meantime, it was vital to hang on by all means possible. His sensitivity to the shifting sands of conflict was doubtless one of the factors that underpinned General Sir David Fraser's assessment that Churchill's 'conceptual reach, at best, far surpassed that of his professional advisers' (Fraser 1999: 93).

Here, it is important to appreciate that 'not losing' can also be a relatively low-cost venture. Indeed, this was exactly what it proved to be for the United States in Afghanistan between 2015 and 2021. In Afghanistan, the

United States was partnered with the Afghan National Defence and Security Forces (ANDSF) which bore the vast brunt of casualties from 2015 onwards, as had been the case since 2001. As well as supplying niche capabilities, often via civilian contractors, the United States played a critical *psychological* role. The mere knowledge that US (and NATO) forces were committed to supporting the ANDSF was crucial to the maintenance of morale and operational capabilities. This was something that the wider Afghan public also understood: in a 2019 survey conducted by the Asia Foundation, 84% of respondents strongly or somewhat agreed that the Afghan National Army needed foreign support to do its job properly (Asia Foundation 2019: 295).

An implication of this psychological role was that any policy initiatives taken by the United States would need to be appraised not just in terms of their immediate military consequences but also of their wider psychological effects since these could translate into military consequences more broadly. The dangers of being seen to 'choose to lose' were recognised by some analysts, but not, it seems, by key policymakers in Washington. This was captured in a study for the Rand Corporation in 2019 in which the authors wrote: 'That there is no military solution to the war in Afghanistan has become a commonplace. But this is, at best, only half true. Winning may not be an available option, but losing certainly is. A precipitous departure, no matter how rationalized, will mean choosing to lose' (Dobbins 2019: 2). And once a major power has been seen as *choosing* to lose, it is only a short step for it to be seen as *being* a loser. The practical consequences of this for its local partner can be profound, ranging from a *sauve qui peut* (general stampede) mentality at the elite level to wholesale panic in the streets. If these eventuate then a war or, more precisely, a particular phase of war can be 'terminated' simply by the collapse of one of the parties. This ultimately was what happened in Afghanistan in August 2021.

How Negotiations Lit The Fuse For Collapse: Cascades

To understand the collapse in Afghanistan, it is useful to draw on a body of literature dealing with social cascades. The fundamental insight underpinning the discussion of cascades is that people can lend support to a cause, or abandon support for a cause, not because of what they would personally like to see happen, but rather on account of what they think is likely to happen given the behaviour of *other actors*. 'As cascades occur', Cass Sunstein has noted, 'beliefs and perspectives spread from some people to others, to the point where many people are relying, not on what they actually know, but on what (they think) other people think' (Sunstein 2009: 90). Perhaps the most spectacular example of cascading in recent decades came with the disintegration of the Eastern Bloc in November-December 1989 (Kuran 1995: 262–288). Once it became clear that the Soviet leadership of Mikhail Gorbachev would not intervene militarily to prop up communist regimes in Eastern Europe, long-suppressed popular detestation of these ruling elites immediately surfaced, beginning with the fall of the Berlin Wall and culminating in the Romanian revolution that overthrew the despised regime of Nicolae Ceauşescu (Siani-Davies 2005).

The phenomenon is, furthermore, one which has appeared in many different contexts. One scholar who has noted its ubiquity is James C Scott, who observed that 'social scientists, not to mention ruling elites, are often taken by surprise by the rapidity with which an apparently deferential, quiescent, and loyal subordinate group is catapulted into mass defiance. That ruling elites should be taken unaware by social eruptions of this kind is due, in part, to the fact that they have been lulled into a false sense of security by the normal posing of the powerless. Neither social scientists nor ruling elites, moreover, are likely to fully appreciate the incitement a successful act of defiance may represent for a subordinate group, precisely because they are unlikely to be much aware of the hidden transcript from which it derives much of its energy' (Scott 1990: 224). A critical point to note, however, is that cascades can be cut two ways. They may occur when people unexpectedly act against a hated regime, but they can also occur when people despairingly align with a toxic force because they have concluded that it is likely to come out on top anyway. Furthermore, it is not necessary that everyone shift at once; rather, at the outset it may only require a small minority to realign themselves for disaster then to ensue, since cascades are cumulative.

What this typically involves is some sort of *trigger*. Once a triggering event occurs, a cascade can develop very quickly, but cascades do not come out of thin air. In the Afghanistan case, two events stand out as relevant, one a trigger, and the other what one might call the addition of accelerant to a fire already burning. The trigger, very obviously, was the Doha Agreement. It was so plainly an exit agreement for the United States rather than a peace agreement for Afghanistan that its propensity to trigger a cascade within Afghanistan should have been obvious. The addition of accelerant came with President Biden's announcement on 14 April 2021 of his intention to follow in all material particulars the commitment of the Trump administration to withdraw all remaining troops and contractors from Afghanistan. At this point, only the most astoundingly naïve could have adhered to the view that there was any more a 'peace process' in place to avert a Taliban takeover. By April 2021, it was clear that the key assumption upon which the Doha Agreement had been based, namely that the Taliban were committed to 'good-faith power-sharing' for the sake of Afghanistan's future, was completely ill-founded.

Is The War Over For Afghans?

The global consequences for the US of its shambolic exit from Afghanistan have been grave, arguably far more serious than those that accompanied the fall of Saigon in April 1975. In part, this reflected weak strategic thinking on the part of the US. As Eliot A Cohen put it, 'U.S. decisions on Afghanistan, Syria, and other trouble spots were … treated as local and separable, with little apparent awareness that they would have global repercussions. It was surely no accident that Russia's annexation of Crimea followed less than a year after the Obama administration failed to enforce its supposed red line on Syria's use of chemical weapons. Nor was it likely a coincidence that Russia invaded Ukraine following the United States' humiliating scuttle from Afghanistan' (Cohen 2022: 124). US officials may comfort themselves that for now, US lives are not directly at risk in Afghanistan, but to the extent that the spectacle of the withdrawal stimulated—and perhaps even triggered—the Russian invasion of Ukraine, the costs of the withdrawal may prove in the long run to have been far higher than anyone in the Biden administration calculated. And if the new Taliban regime provides an enabling environment for *al-Qaeda*, the US could discover that washing its hands of Afghanistan was not quite as easy as it might have seemed.

'Any man's death diminishes me, because I am involved in mankind. And therefore never send to know for whom the bell tolls: it tolls for thee', so wrote John Donne in 1623 (quoted in Scott 1997: 75). After the US exit from Afghanistan, there was little disposition in Washington to listen to the bells tolling for the Afghans, but it is important in conclusion to address one fundamental question: is the war over for *them*? Recent history in Afghanistan is not especially encouraging. The 1989 Geneva Accords on Afghanistan, spruiked as giving effect to Afghans' right to self-determination, simply remitted to the battlefield the underlying conflict that the Accords had failed to address (Maley 1989). The end of US involvement in Afghanistan may not presage a flare-up on the battlefield, but it would be a serious mistake to treat it as bringing 'peace' for the Afghans. Such a line of argument would be akin to suggesting that Nazi Germany brought 'peace' to Poland when its forces overran Warsaw in late September 1939. The underlying and profound conflict in values between the Taliban and the bulk of the Afghan population remains entirely unresolved. The Asia Foundation in its 2019 survey found that 85.1% of respondents had 'no sympathy at all' for the Taliban (Asia Foundation 2019: 19, 315), an unsurprising finding given the youth of the Afghan population and its exposure to more than two decades of globalisation. In the light of this chasm in values, the risk to the Taliban regime of popular protests in urban areas—a manifestation of 'contentious politics'—is likely very high.

With such low levels of popularity, the likelihood that the Taliban regime would be able to secure generalised normative support is low. Likewise, the enfeeblement of the Afghan state as a result of the loss of critical personnel and of foreign funds to bolster state activities means that the capacity of the Taliban to *buy* support is also limited. As a consequence, there is every reason to fear that the Taliban regime might embark on an even more aggressive campaign of coercion to try to secure its position. This, however, could prove self-defeating: the use of state terror by the communist regime during the 1980s had a boomerang effect, driving more and more people into opposition (Maley 1991); and disgruntled Taliban figures might break away on ethnic lines, creating new challenges for the regime to address.

There is one final danger that haunts at least some of Afghanistan's people: the danger of genocide. This has received some attention in the context of vicious bombings in Kabul and elsewhere directed at Afghanistan's Hazara minority (Genocide Watch 2022), which has a long history of being marginalised or persecuted (Ibrahimi 2017). Genocide in this context has a precise meaning, set out in Article II of the *Convention on the Prevention and Punishment of the Crime of Genocide* of 1948. Article II defines as genocide 'acts committed with intent to destroy, in whole or in part, a national, ethnical, racial or religious group', and includes causing 'serious bodily or mental harm to members of the group' and 'Deliberately inflicting on the group conditions of life calculated to bring about its physical destruction in whole or in part'. While responsibility for some attacks has been claimed by Islamic State, the distance between the Taliban and Islamic State is not as great as some optimists would like to think (Ibrahimi 2020). Taliban have a record of conducting atrocities targeting Hazaras; in August 1998 in Mazar-e Sharif they carried out a massacre that the writer Ahmed Rashid described as 'genocidal in its ferocity' (Rashid 2000: 73).

In 2003, one of the leading specialists on genocide, Barbara Harff, put forward a six-variable model for assessing risks of genocide and mass political murder, the variables being (1) magnitude of prior political upheaval; (2) experience of prior genocides or politicides; (3) exclusionary elite ideology; (4) autocratic regime type; (5) an elite based mainly or entirely on an ethnic minority; and (6) international interdependence (Harff 2003). Afghanistan under the Taliban meets all of these criteria. And even before the Taliban takeover, analysis of data from the Atrocity Forecasting Project at The Australian National University (see Butcher 2020: 1524–1547) had placed Afghanistan in the top five countries in the world at risk of genocide or 'politicide' in 2021–2023. While such forecasts are probabilistic rather than definitive, they should prompt very sobering reflections on where Taliban domination could lead. Wars can and do end, and in different ways (Taylor 1985; Iklé 1991), but for the people of Afghanistan, the notion that war has come to an end seems tragically premature.

References

Alexander, C., 2021. *Ending Pakistan's Proxy War in Afghanistan*. Macdonald-Laurier Institute, Ottawa.

Asia Foundation, 2019. *A Survey of the Afghan People: Afghanistan in 2019*. The Asia Foundation, Kabul.

Butcher, C., Goldsmith, B.E., Nanlohy, S., Sowmya, A. and Muchlinski, D., 2020. 'Introducing the Targeted Mass Killing Data Set for the Study and Forecasting of Mass Atrocities', *Journal of Conflict Resolution*, vol. 64, nos 7-8.

Chong, D. and Druckman J.N., 2007. 'Framing Theory', *Annual Review of Political Science*, vol. 10, pp 103–126.
Clark, K., 2021a. *The Taleban's Rise to Power: As the US Prepared for Peace, the Taleban Prepared for War*. Afghanistan Analysts Network, Kabul.
Clark, K., 2021b. *Afghanistan's Conflict in 2021 (2): Republic Collapse and Taleban Victory in the Long View of History*. Afghanistan Analysts Network, Kabul.
Cohen, E.A., 2022. 'The Return of Statecraft: Back to Basics in the Post-American World', *Foreign Affairs*, vol. 101, no. 3, pp 117–129.
Desai, R. and Eckstein H., 1990. 'Insurgency: The Transformation of Peasant Rebellion', *World Politics*, vol. 42, no. 4, pp 441–465.
Dobbins, J., Campbell, J.H., Mann, S. and Miller, L.E., 2019. *Consequences of a Precipitous U.S. Withdrawal from Afghanistan*. RAND Corporation, Santa Monica.
Eikenberry K.W., 2009. 'COIN Strategy: Civilian Concerns'. *U.S. Department of State Cable no 03572*, Kabul.
Farasoo, A., 2021. 'Rethinking Proxy War Theory in IR: A Critical Analysis of Principal–Agent Theory', *International Studies Review*, vol. 23, no. 4, pp 1835–1858.
Feynman, R.P., 2001, *The Pleasure of Finding Things Out*. Penguin Books, London.
Fraser, D., 1999. 'Alanbrooke', in John Keegan (ed.), *Churchill's Generals*. Abacus, London, pp 89–103.
Genocide Watch, 2022. *Genocide Emergency: Afghanistan*. Genocide Watch, Washington DC.
Haqqani, H. and Curtis, L., 2017. *A New U.S. Approach to Pakistan: Enforcing Aid Conditions without Cutting Ties*. Briefing Paper, The Hudson Institute, Washington DC.
Harff, B., 2003. 'No Lessons Learned from the Holocaust? Assessing Risks of Genocide and Political Mass Murder since 1955', *American Political Science Review*, vol. 97, no. 1, pp. 57–73.
Hobbes, T., 1996. *Leviathan*. Cambridge University Press, Cambridge.
Ibrahimi, N., 2017. *The Hazaras and the Afghan State: Rebellion, Exclusion and the Struggle for Recognition*, Hurst & Co, London.
Ibrahimi, N. and Akbarzadeh, S., 2020. 'Intra-*Jihadist* Conflict and Cooperation: Islamic State–Khorasan Province and the Taliban in Afghanistan', *Studies in Conflict and Terrorism*, vol. 43, no. 12, pp. 1086–1107.
Ibrahimi, N. and Maley, W., 2020. *Afghanistan: Politics and Economics in a Globalising State*. Routledge, London.
Iklé, F.C., 1991. *Every War Must End*. Columbia University Press, New York.
Jamal, A.S. and Maley, W., 2022. *The Decline and Fall of Republican Afghanistan*. Hurst & Co, London.
Kaldor, M., 2012. *New and Old Wars: Organized Violence in a Global Era*. Stanford University Press, Stanford.
Kilcullen, D., 2009. *The Accidental Guerrilla: Fighting Small Wars in the Midst of a Big One*. Oxford University Press, New York.
Kilcullen, D., 2010. *Counterinsurgency*. Oxford University Press, New York.
Kuran, T., 1995. *Private Truths, Public Lies: The Social Consequences of Preference Falsification*. Harvard University Press, Cambridge.

Lukacs, J., 1999. *Five Days in London: May 1940*. Yale University Press, New Haven.

Maley, W., 1989. 'The Geneva Accords of April 1988', in Saikal, A., and Maley, W., (eds.), *The Soviet Withdrawal from Afghanistan*. Cambridge University Press, Cambridge, pp. 12–28.

Maley, W., 1991. 'Social Dynamics and the Disutility of Terror: Afghanistan, 1978-1989', in Bushnell, P.T., Shlapentokh, V., Vanderpool, C.K. and Sundram, J. (eds.), *State Organized Terror: The Case of Violent Internal Repression*. Westview Press, Boulder CO, pp. 113–131.

Maley, W., 2002. 'Confronting Creeping Invasions: Afghanistan, the UN and the World Community', in Warikoo, K. (ed.), *The Afghanistan Crisis: Issues and Perspectives*. Bhavana Books, New Delhi, pp. 256–274.

Maley, W., 2016. *Transitioning from Military Interventions to Long-Term Counter-Terrorism Policy: The Case of Afghanistan (2001-2016)*. International Centre for Counter-Terrorism, The Hague.

Maley, W., 2018. *Transition in Afghanistan: Hope, Despair and the Limits of Statebuilding*. Routledge, London.

Maley, W., 2021a, *The Afghanistan Wars*. Macmillan/Red Globe Press, London.

Maley, W., 2021b. 'Terrorism and insurgency in Afghanistan', in Izarali, M.R. and Ahlawat, D. (eds), *Terrorism, Security and Development in South Asia: National, Regional and Global Implications*. Routledge, London. pp. 140–156.

Maley, W. and Jamal A.S., 2022. 'Diplomacy of disaster: The Afghanistan 'peace process' and the Taliban occupation of Kabul', *The Hague Journal of Diplomacy*, vol. 17, no. 1, pp 32–63.

Orvis, K.A., 2021. *Department of Defence (DoD) Quarterly Suicide Report (QSR) 2nd Quarter, CY2021*. Defense Suicide Prevention Office, Department of Defense, Washington DC.

Rashid, A., 2000. *Taliban: Militant Islam, Oil and Fundamentalism in Central Asia*. Yale University Press, New Haven.

Scott, J.C., 1990. *Domination and the Arts of Resistance: Hidden Transcripts*. Yale University Press, New Haven.

Scott, R., 1997. *No man is an island: A selection from the prose of John Donne*. The Folio Society, London.

Sharan, T., 2023. *Inside Afghanistan: Political Networks, Informal Order, and State Disruption*. Routledge, London.

Siani-Davies, P., 2005. *The Romanian Revolution of December 1989*. Cornell University Press, Ithaca.

Sunstein, C.R., 2009. *Going to Extremes: How Like Minds Unite and Divide*. Oxford University Press, New York.

Taylor, A.J.P., 1985. *How Wars End*. Hamish Hamilton, London.

Weinbaum, M.G. and Harder, J.B., 2008. 'Pakistan's Afghan policies and their consequences', *Contemporary South Asia*, vol. 16, no. 1, pp. 25–38.

Section 3
Alternatives

This section examines what might be seen as alternatives to traditional solutions to ending wars.

John Blaxland explores the Australian experience since World War II and examines the relationship between military, diplomatic and economic levers of power in the ending of conflict, demonstrating that military intervention is only part of the toolkit in successful conflict resolution. Given the rise in great power competition, the impact of climate change (in particular in the South Pacific) and a contest in political systems, it makes sense to invest further in all aspects of the conflict resolution toolkit to meet the challenges of tomorrow.

All recent wars have, thankfully, been between states that are not both nuclear-armed major powers. We therefore do not have a repository of experience of how such wars between major powers might progress or end. Instead, we rely on nuclear deterrence theory developed during the Cold War in a bi-polar environment where, for both the USA and USSR, the stakes were broadly equal. Yet, the world now has a wider range of nuclear-armed and ambitious states, each of whom has its own objectives that might not necessarily be of equal importance to a potential adversary—thereby undermining a key element of deterrence. Hugh White examines how great power war might evolve in two scenarios: one in Europe between the USA and Russia; the second in Asia between the USA and China. He concludes that, in each case, the USA may not be prepared to escalate to use nuclear weapons and, as a result, its nuclear force may no longer be sufficient to deter major power war.

DOI: 10.4324/9781003317487-13

10 Reflections on The Australian Experience: How Wars End[1]

John Blaxland

Looking Back

At the height of the Second World War, Australia, a country of under eight million people, mustered land, air and maritime forces at an industrial scale. This included an army of fourteen divisions—with each division incorporating multiple brigades and subordinate battalions of infantry armoured and motorised formations (Palazzo 2001). The Royal Australian Air Force (RAAF) peaked in late 1944 at over 182,000 personnel and 6,200 aircraft in 61 squadrons. In 1945, it is said, Australia had the fourth-largest air force in the world (after the USA, USSR and UK) when counting operational aircraft squadrons (including fighters, transport, bombers and reconnaissance aircraft) (Odgers 1957: 5).

Before the end of the war the peak strength of the Royal Australian Navy (RAN) had reached 39,650 with approximately 337 vessels ranging from cruisers to motor launches in service (Pelvin undated). Experience with the British Pacific Fleet which operated from Australian ports from early 1945 onwards, exposed the RAN to Britain's latest warships including multiple aircraft carriers—a sight which helped inspire Australia's post-war choice to acquire two for the RAN, HMAS *Sydney* and *Melbourne*. This was an armed force built for an industrial-era war. The unconditional surrender of Germany and Japan appeared to vindicate that approach. Meanwhile, as the war's end approached, the United States offered to host the establishment of the United Nations—an entity that would seek to circumscribe the use of war as an instrument of state. Thereafter a body of international precedents and international legal principles emerged which established de facto limits on the exercise of military power. Those limits were not absolute, except perhaps, in practice (so far at least) concerning the use of nuclear weapons in anger.

After The World Wars

Indeed, the ending of the war with the dropping of two atomic bombs on Hiroshima and Nagasaki pointed to a very different set of circumstances that would follow. Within four years, the Soviet Union also had tested its

DOI: 10.4324/9781003317487-14

own nuclear weapons and the era of mutually assured destruction followed, whereby the contest between the Soviet bloc and the US-aligned West was conducted largely through proxies. For Australia, this was most notably the case with the Korean War and Vietnam War. No longer would wars be settled purely as a result of martial power. Other elements of national power would be wielded to generate the policy effects that previously had relied on soldiers, guns, tanks, bombers and warships to achieve.

In *Politics Among Nations* (1948) Hans Morgenthau outlined a proposal that identified nine elements of national power. These were geography, resources, industrial capacity, military preparedness, population, national character, national morale, quality of diplomacy and quality of government. His work pointed to the broader range of instrumentalities to be wielded in the pursuit of politics and a nation's interests. The post-war atomic era, with the UN framework as a supporting trestle, saw major conflict on the scale of the world wars avoided for over 75 years and various elements of national power wielded for political effect. While the doomsday clock, they say, is stuck at the eleventh hour only minutes from 'midnight', thankfully no country has employed nuclear weapons against another since then.

Australia was a founding member of the United Nations, but fearful of a repeat of the events a decade earlier, signed a security treaty with the United States and New Zealand in 1951. The ANZUS Treaty, aligning Australia with the world's largest economy, most powerful armed force and leading democracy helped reinforce a sense of confidence that Australian defence expenditure could be constrained. This restraint was enabled by three factors. First, the combination of the international order, represented by the establishment of the United Nations. Second, the nuclear stalemate, with mutually assured destruction effectively restrained potential great power excess on the battlefield. And third was the alliance arrangements, which were represented most visibly by the ANZUS Treaty, but followed by other security mechanisms, including the establishment of the Joint Defence Facility at Pine Gap. In light of this overlapping triangle of factors, no further need was seen for a major industrial-scale armed force to defend Australia. Operations of necessity, such as the defence of Australia in 1942 onwards, now seemed a remote prospect. Instead, at most, Australia would be called upon to make contributions to military operations of choice—most of which would arise far from Australia's shores.

Instead of the earlier industrial scale mobilisation of force, a boutique defence force would emerge centred on a fleet of about 100 fighter aircraft and supporting air platforms, a dozen or so warships and a handful of submarines, coupled with a professional (full-time) army centred on what would settle at about three brigades in one infantry division, supported by a handful of reserve (militia) formations (brigades with subordinate battalions) in a second, but relatively hollow, infantry division.

There were fluctuations around the edges, but this force outline sustained Australia's military commitments in the operations it chose to become involve with, such as the Korean War and the Vietnam War, its deployment of forces to support the British Commonwealth Far Eastern Strategic Reserve, based in Malaysia, and its support for British and Malaysian efforts to contain Indonesian incursions which materialised as part of President Sukarno's short-lived *Konfrontasi* (Edwards 1992). In these instances, Australia sought to make a substantive but carefully calibrated contribution far short of the scale of engagement mustered during the preceding world wars. That was considered adequate because of the limited nature of the wars involved (due in part to the balance of power and UN superstructure), as well as the overwhelming commitment of the United States (in Korea and Vietnam) and United Kingdom (in Malaysia) which preclude the need for a more muscular contribution from Australia.

In each of these instances, the military contribution was widely regarded as significant, but not the deciding factor, in the wars' outcomes. In Korea, the stalemate and reluctance to cross the nuclear threshold meant political compromise was key to reaching an armistice. The war ended in practice, but technically continues to this day. In a similar manner, Vietnam, exhausted US political will, but ended more categorically with the South Vietnamese forces routed in April 1975. The North Vietnamese, with Soviet backing, had more resolve, more 'skin in the game' than did the United States. That meant that battlefield successes mattered little in the context where the politics were not reconciled with the strategy.

In the case of Indonesia's *Konfrontasi,* carefully calibrated and secretive combined British, Australian, Malaysian and New Zealand efforts along the Indonesian border, enabled with signals intelligence insights, stymied Sukarno's initiatives. Eventually though, it was the overthrow of Sukarno by his successor Suharto, and the change of direction of domestic Indonesian politics that followed which dissipated Indonesian enthusiasm to continue its incursions into Malaysian territory. Military force helped, yes, but domestic politics in Jakarta was the key determinant.

The Post-Cold War Period: Far Away and Closer to Home

The post-Cold war era seemed to point to the 'end of history' or the end of the dialectical contest between competing political systems. (Fukuyama 1989). The Gulf War in 1991 further seemed to demonstrate the United States had learnt from its Vietnam mistakes militarily and emerged dominant strategically. That war ended categorically and convincingly, it seemed, but laid the seeds for the subsequent US-led invasion in 2003—a conflict which in turn spawned a series of violent political and military convulsions across and beyond Iraq, the likes of which are still being dealt with today.

Meanwhile, the American unipolar moment that followed appeared to vindicate their so-called 'Revolution in Military Affairs' which stressed precision-guided munitions and technological overmatch to defeat adversaries and cower would-be challengers. In this context, Australia comfortably supported other operations of choice during the 1990s including UN-endorsed peacekeeping, humanitarian assistance and disaster relief operations in Africa and Asia (Horner 2011, 2014). Many of these peacekeeping operations were the final acts in longstanding wars that had been enduring for years during the Cold War; or they were more warlike manifestations of conflict that remained unresolved.

In many of these conflict situations, or post-conflict situations that remained in the twilight zone between peace and war, Australia's armed forces found themselves working closely with a range of agencies and organisations. Many of these are ones that Morgenthau would have identified as being among the nine elements of power. These include the International Committee for the Red Cross/Crescent, the World Food Program, *Medicins Sans Frontieres,* World Vision and a range of other international aid agencies. In addition, Australia operated an aid and development program through what is now known as Australian Aid but previously known as AusAid and before that, as the Australian International Development Assistance Bureau (AIDAB). Whereas in earlier wars, Australian armed forces had operated alongside allied counterparts, in the post-Cold war years, this happened again, but more frequently alongside other Australian government agencies including Ausaid, the Australian Federal Police (AFP) and diplomats from the Department of Foreign Affairs and Trade (DFAT).

In Namibia in 1989–1991, for instance, Australia's contribution of an engineering task force and headquarters staff as part of an UN-mandated peacekeeping mission was instrumental in resolving a crisis that had run its course but which needed support to fully resolve. Diplomacy, complemented by measured application of military capabilities generated a positive outcome. Namibia then was able to join the family of nations as a fully-fledged UN member state.

In Cambodia in 1992-94, Australia's deft diplomacy, resolute engagement and carefully calibrated contribution of headquarters staff, advisors, communicators and aviation teams helped facilitate the rehabilitation of Cambodia into the family of nations after years of war. This involved a high level of collaboration and coordination with a range of UN-related and other aid and development international organisations. Arguably, however, the lack of resolve over the ballot set the path for Hun Sen's authoritarian model of governance to emerge.

In Somalia, in 1993-94, Australia offered 'a little bit of hope' (Breen 1998) with the deployment on Australia's naval amphibious ships of a force centred on the 1st Battalion of the Royal Australian (Infantry) Regiment, commanded by then Lieutenant Colonel (now Governor General) David Hurley; but the Australian government's contribution was a niche one, with little thought given to the longer term ramifications on the impoverished country or on what that would mean for Australia. The federal government in Canberra did not consider it was in a position to make a major contribution to making a lasting difference there. In hindsight, there is little evidence left there of Australia's efforts. Instead, a thriving migrant Somali community has established itself in Australia.

In Rwanda from 1993 to 1996, Australia contributed to an UN-mandated assistance mission that helped the country find its feet after a harrowing attempted genocide. Two rotations of a composite medical-military team deployed there to assist in the nation's recovery. There they worked alongside national contingents supplied to the United Nations from around the world, as well as a range of humanitarian assistance agencies. Yet without a mandate to stop the violence, the experience was as traumatic for Australian soldiers as it was helpful to the future of Rwanda. Peace was restored, technically, but a fraught one.

In the island territory of Papua New Guinea known as Bougainville, Australia intervened briefly in 1994 and again from 1998–2001 (Breen 2016). This intervention involved unarmed Australian military peacekeepers, working alongside New Zealand Defence Force personnel and a combined group of civilian diplomats and aid workers and contingents from Fiji and parts of the Pacific. Here the international collaboration from Pacific neighbours proved of more than passing utility. They were tasked to monitor a truce and subsequent peace agreement following over a decade of strife arising from disaffection over mishandling of the Panguna copper mine in the centre of the island. This approach, involving respectful and culturally-attuned inter-agency and international engagement, helped resolve the dispute and set Bougainville on a path to autonomy if not outright independence. The military presence was a significant enabler with security, logistics, communications, health and other support. But the military presence alone was insufficient. The combined effect of diplomacy, aid, education and political compromise facilitated in Bougainville and Port Moresby was key.

Several elements of national power that Morgenthau identified came into play in this intervention. Notably in terms of geography, Australia's proximity and interest was an important factor in the attention the Bougainville conflict received in Canberra. Bougainville is resource-rich. The mishandling of those resources had contributed directly to the conflict which the truce and peace monitoring group sought to address. Military preparedness was tested for Australia, as well as for the international contingents from the Pacific region that Australia sought to coordinate, but proved able to deliver on the needs on this occasion. In terms of national character this engagement as a responsible Pacific neighbour was widely appreciated not just in Bougainville but in the rest of Papua New Guinea and further afield. In terms of national morale, this event bolstered confidence in the image of the Australian soldiers as 'soldier, teacher, ambassador, peacekeeper'. The quality of diplomacy speaks for itself—including the work of diplomats on the ground in Bougainville and in the capitals of the participating Pacific nations. In terms of quality of government, the work in Bougainville sought to model the liberal democratic way. Over two decades later, the interim assessment is that it has worked to this point.

Today the legacy of that intervention unfolds as Bougainville strives to assert its independence by non-violent means.

Australia's collaborative experience in Bougainville serendipitously helped it prepare for an even greater challenge in 1999, following Indonesian President Habibie's decision to conduct a plebiscite about the future of East Timor. Australia and a range of countries convinced Indonesia to support an unarmed United Nations sponsored Assistance Mission to East Timor (UNAMET) to monitor the ballot. As pundits predicted, the outcome favoured independence and triggered Indonesian-sponsored militias to wreak havoc in revenge. Indeed, the structure of the Indonesian military, the TNI, and their militias, always focused on dominating an essentially civilian population. They were unprepared when they confronted an actual organised army presence (Kingsbury 2009).

In responding, the international community reacted in shock and promptly supported an UN-sanctioned and Australian-led International Force in East Timor (INTERFET). This widely televised crisis saw military force on display, but there were other factors at play behind the scenes. The INTERFET commander then Major General (later Governor General) Peter Cosgrove masterfully used the media to aid his cause. But despite the operation being one that saw very few casualties once the force deployed, he later opined that a 'less robust force optimised for peacekeeping would have invited more adventurist behaviours by our adversaries'. (Blaxland 2002). Cosgrove makes the point that the military had a role to play—and from the author's personal experience, the first 48–72 hours did seem to indicate that a deadly armed clash was not something to be ruled out as implausible or unlikely. But what other factors were involved and to what extent was a military solution relevant and necessary here? Economic coercion, notably from the United States, and Australian diplomacy and regional engagement certainly were important catalysts.

First is economic influences. The US President Bill Clinton was reluctant to commit to US 'boots on the ground', having a range of other security challenges in mind, not the least of which was the conflict in Kosovo ongoing at the time. Yet, with prompting from his Commander Pacific Command, Admiral Dennis Blair, Clinton came to appreciate the need to be supportive of his Australian ally. At the APEC summit in Auckland in early September, Clinton engaged with Habibie, Howards and others. Habibie and the Indonesian military leaders came to appreciate that, particularly through the IMF, the United States had economic levers it could wield to pressure Indonesia into accepting international intervention. The threat of economic sanctions shortly after the already-crippling Asian financial crisis of 1997–1998 would have had a debilitating effect on the Indonesian economy. This seems to have been a significant factor in convincing Indonesia to accept the Australian-led international force and back down.

Second is the place of diplomacy and regional engagement. Initially US efforts to maintain a different stance towards Indonesia from that of

Australia caused considerable consternation with Prime Minister John Howard surprised and disappointed at the reluctance to commit ground forces. In the end, they were not required. What was required, though, was US military transport and logistic support to help deploy and sustain a 22-country coalition in a small country that had just been ravaged by rampaging pro-Indonesia forces. In hindsight, it worked in Australia's interests for the United States to take a different approach, to not be too closely associated with Australia's position or be seen to be overly beholden to Australia's approach. Besides, with the behind-the-scenes support provided, Australia did not end up needing a substantial US military contribution. Diplomacy, some coercion, calling in favours and a bit of bluff all contributed.

In addition, while working out how to manage the fallout, Indonesia sought an ASEAN member state to offer to lead the intervention force. No one volunteered other than Australia. In doing so, however, Australia assiduously called on favours with Southeast Asian partner nations, sending the Vice Chief of Defence Force, Air Marshal Doug Riding, on a tour around the region soliciting support. Initially the VCDF approached the closest ones with security links to Australia, which were Singapore and Malaysia—notably tied to Australia (along with New Zealand and the United Kingdom) through the Five Power Defence Arrangements (FPDA). But being immediate neighbours to Indonesia, they were understandably reluctant to be the first among ASEAN states to step forward. One step removed, but still with close bilateral ties to Australia, Thailand offered to appoint a deputy force commander and a substantial task force to support. This broke the logjam, and the Philippines, Malaysia and Singapore soon followed with offers of contingents. This work reflected decades of investment in bilateral and multilateral ties—through the efforts of Australian diplomats and through Australia's military attaché network across the region. It is the moment which most clearly validated Australia's investment in regional defence and security ties.

New Zealand, which had been such a crucial partner to the truce and peace monitoring in Bougainville, once again stepped forward, providing air, naval and ground forces in support—including the 1st Battalion of the Royal New Zealand (Infantry) Regiment. As host of the APEC summit days earlier, New Zealand had also played a pivotal facilitator role for tense but crucial face-to-face negotiations and deliberations.

What is clear from Australia's involvement in the Bougainville and East Timor crisis is the key role of diplomacy, backed up with credible but restrained military capabilities and supported by regional engagement with coalition partners and the provision of humanitarian assistance. Morgenthau's nine elements of power (geography, resources, industrial capacity, military preparedness, population, national character, national morale, quality of diplomacy, and quality of government) were on clear display.

In these circumstances, Australia took the 'primary risks and primary responsibilities' (Ryan 2000), thinking through the issues, engaging partners and planning ahead. A range of government instrumentalities was activated, an East Timor task force was established, inter-departmental task forces were invoked and collaborative and relatively collegial planning and management came to the fore (Connery 2010). Without doubt, more could have been done and what was done could have been done better. But Australia at least recognized that in stepping up to lead in these crises, it needed to bring in a range of government capabilities and not rely solely on the armed forces. In the end, it is fair to say that it was this unique combination that resulted in what pundits described as 'mission accomplished' (Breen 2000). Yet even here, results were marred by incomplete political and social reconstruction—such that Timorese society was marred by an outbreak of violence in May 2006—one that reflected unresolved disputes that the Australian-led intervention pasted over in its effort to dis-engage and facilitate independence.

Since 9/11

Then along came 9/11. Prime Minister John Howard was in Washington for the 50th anniversary celebrations of the signing of the ANZUS Treaty. Being moved by what he saw, heard and felt while in Washington on that fateful day, he came to see it as cause to invoke the treaty and to call for an Australian military response in support of the US-led efforts. (Blaxland 2020)

The problematically named 'Global War on Terror', saw Afghanistan feature as the haven of the terrorist group behind the 9/11 attacks—Al Qaeda. A military solution was applied, relying heavily on US special forces in conjunction with co-opted Afghan forces form the so-called Northern Alliance. Initially this was surprisingly successful. Bolstered by his own judgement and early successes, US Defense Secretary Donald Rumsfeld refused to deploy sufficient forces to defeat in detail Al Qaeda and its Taliban supporters. Instead, he chose to co-opt additional local forces for a cordon and sweep operation. That operation was intended to defeat Al Qaeda in the Tora Bora Mountains on the Pakistan border, but it failed in part because insufficient force was applied and the locally recruited forces, in turn, were co-opted back by the Taliban, facilitating their escape across the border.

Notwithstanding the incomplete defeat of Al Qaeda, reconstruction and rehabilitation of Afghanistan seemed eminently achievable in the deliberations of 2002. Eager to retract its forces promptly, Australia took its lead from the Americans on the strategy for managing post-Taliban governance. This was not a domain Australian policymakers were encouraged or inclined to seek to steer closely—in marked contrast to the approach taken only a few months earlier to the management of the East Timor crisis. Afghanistan, after all, was an operation of choice far from Australia's shores. East Timor, in contrast, had seemed more of an operation of necessity and much closer to home. Little did they realise how half-baked plans imposed by well-meaning but culturally discordant and largely historically ignorant Western outsiders would so gravely become derailed in short order.

Meanwhile with the United States eager to pursue illusory weapons of mass destruction in Iraq, Australia once again signed up in support, but left the strategy formulation and planning to their American allies, mindful that, once again, this was an operation of choice and far from Australia's shores. Howard reckoned that a carefully-calibrated niche-force contribution would satisfy alliance requirements and leave Australia free to return its forces to barracks in short order. The fiasco (Ricks 2007) that unfolded thereafter reflects the failure to understand the local cultural, political, religious, demographic facts of a divided and repressed nation like Iraq; let alone the effects that dismantling the instruments of state that had held the nation together would have in fuelling the flames of instability, insurgency and civil war.

While there was much to criticise about the repressive regime of Saddam Hussein, military force was only ever likely to be a portion of any solution to the problems of Iraq. Yet the military was the principal instrument of state wielded to shape the post-Saddam order—all without being adequately informed by the other factors and with a US military approach which prided itself on its 'warfighter' reputation that shunned tasks seen as of a lesser order of importance like peacekeeping and nation building.

In the Iraq War, it is evident that Morgenthau's nine elements of power were applied inconsistently. While many factors were at play, a heavy emphasis on US military preparedness mean that not enough emphasis was placed on the other elements of national power including geography, resources, industrial capacity, population, national character, national morale, quality of diplomacy, and quality of government.

The collapse of the Islamic Republic of Afghanistan in August 2021 in the face of Taliban advances shocked many with its speed. The outcome, which echoed the fall of US partner South Vietnam in the face of a North Vietnamese invasion a generation earlier, pointed to the criticality of political will. The United States having not been defeated in its military engagements at the tactical and operational levels, pursued a strategy, supported by Australia and other coalition partners that was set on a course for defeat in what Karen Middleton described as *An Unwinnable War* (Middleton 2011).

Once again key elements of national power affecting the outcome in Afghanistan were not sufficiently understood or given sufficient weight. While exploring these issues in full would take another full treatise, there are some brief observations worth making, starting with geography.

The geography of Afghanistan was problematic. Being a landlocked country surrounded by Pakistan, Iran and the former soviet republics, the 'stans' to the north, meant accessing and operating inside Afghanistan was more challenging. Dependence on collaboration with Pakistan was deeply problematic because of Pakistan's mixed allegiances and fear of Western intervention.

Resources were never managed well. Local Afghanistan economic resources were poorly managed and exploited by the Taliban. The people were pulled in different directions by competing visions, let alone threats and risks. The national character, if it is fair to characterise it, was one Western forces were too slow to come to understand. Arguably a succession of six to twelve-month campaigns was prosecuted by well-meaning but arrogant and largely ignorant outsiders compounded rather than solved a series of governance challenges. Morale not surprisingly plummeted when the US no longer was prepared to act as the state's guarantor.

In the meantime, regional challenges relating to storms, tsunamis, earthquakes and breakdowns in law and order have generated a wave of demands for rapid deployment of ADF elements to provide emergency assistance. Enhanced regional collaboration on humanitarian assistance and disaster relief is the silver lining to this dark cloud. Collaboration has manifested in regional bodies like the Pacific Island Forum and the ASEAN Defence Ministers Meeting-Plus (ADMM-Plus) construct, where expert working groups have grappled with issues ranging from counter-terrorism, piracy and maritime security, military medicine, peacekeeping operations, humanitarian mine action and cyber security.

Future Overlaps and Conundrums (Blaxland 2019)

At the time of writing, Australia faces challenges relating to an overlapping range of issues spanning great power contestation, looming environmental catastrophe and a spectrum of governance challenges (notably related to terrorism, smuggling and a breakdown in law and order), particularly in some of the smaller and more vulnerable states in the Pacific. These are compounded by developments related to the fourth industrial revolution, leveraging off robotics, artificial intelligence, quantum computing and the so-called internet of things. Military power has demonstrated utility in addressing various facets of these overlapping challenges but is not a solution, in and of itself, to any of them.

For great power contestation, having a credible force to deter would-be aggressors and protect one's national interests is broadly recognised as a key responsibility of the nation state. The devil is in the detail of formulating and sustaining a suite of capabilities that offer reassurance and reduced the prospect of a potential adversary seeking to exploit one's weaknesses—all without bankrupting the nation.

With looming environmental catastrophe, the armed forces are called upon increasingly frequently as a stop-gap measure to face fires, floods, storm and pestilence (including pandemics). This demonstrates the remarkable agility and adaptability of the men and women in Australia's Defence Force. For a boutique defence force, managing these crises consecutively has proven difficult but manageable for now. Should they start coming more frequently or concurrently, the ADF will be hard-pressed to handle them.

Then there is the spectrum of governance challenges for which the military can often get called upon to assist. This includes a range of terrorist-related incidents, people smuggling, drug trafficking and associated corruption contributing to a breakdown in law and order—at home and abroad.

In addition, there is increasing prevalence of technology in society, aiding and at times replacing the human workforce with artificial intelligence, robotics, quantum computing, persistent surveillance and unprecedentedly precise tracking of movement, actions and thoughts.

With so many factors interacting, the old idea of war, for many at least, seems to have been replaced by the idea of grey zone or hybrid operations—although admittedly Putin's invasion of Ukraine has put conventional offensive military operations back in the news. Even then, though, the tactics employed involve a range of instrumentalities and domains, notably including cyber-attacks and media message manipulation.

Today, it may be best to describe interstate relations along a spectrum ranging at one end from cooperation to competition to contestation to conflict in which Morgenthau's nine elements of national power are exercised routinely. Apart from very rare interstate armed conflicts, Australia, like other nations, is faced with more 'small wars' variously defined. This then requires a different approach to purely conventional military responses, and the inclusion of non-military solutions.

A Way Forward

This cooperation to conflict continuum presents significant challenges. In and of themselves, each of the broad categories identified fall along that continuum and presents challenges for which easy solutions are hard to find. When they are aggregated, however, they present problems beyond the remit of any one government agency, state, federal or international jurisdiction, let alone any academic discipline or skill set.

When considering the efficacy of the use of force into the future, the limitations of armed forces' capabilities should be front and centre in our thinking. Now, though, that also needs to be set in the context of overlapping challenges that require a re-imagination of national security and of shared concerns affecting all of humanity.

For many of these challenges, the military has a supporting but not necessarily direct or primary role. Yet the ADF continues to be called upon to play prominent roles in the light of floods, fires, pandemics, cyber-attacks, international crises (human-centric and other disasters) and domestic security challenges.

Much of this reflects policy decision-making that has been undertaken on the run, responding to the tyranny of the urgent and the 24/7 news cycle. Leaders can barely catch their breath and think through the ramifications of decisions made with short-term challenges in mind. Policymakers have little time or space to reflect more deeply on the trends and on pointers for satisfactory solutions to problems appearing on the horizon.

In reflecting on the way forward, reconsidering Morgenthau's nine elements of national power and their role in ending wars or mitigating the risks and the effects of wars, is timely.

With this in mind, several initiatives have been launched including the creation of the Institute for Climate, Emergency and Disaster Solutions (ICEDS) at the Australian National University. Other campuses have similar inter-disciplinary programs to help focus research in response to some of the bigger existential issues faced by contemporary society. I would contend the government should take this idea and go further, helping to establish a national institute for net assessment (NINA). Such an institute could help to think through the challenges Australia faces in a holistic and interdisciplinary manner, undertaking a thorough assessment of the risks and opportunities faced, of what the future holds, and how best to position for the challenges ahead.

More specifically, in terms of capabilities to respond to the emerging conflagration of crises, the boutique defence force, designed for the post-Cold War unipolar moment is arguably no longer fit for purpose. It seems that this has spurred further investment in the Defence Strategic Update of July 2020 and the September 2021 announcement of the AUKUS arrangement to supply Australia with nuclear propulsion submarines and a suite of precision longer-range missile systems to bolster the ADF's ability to deter and, if necessary, strike an adversary. This muscling up of the defence force is not going to be enough if that continuum and the overlapping challenges continue to stretch our ability to respond.

In addition, other agencies need a refresh, including at local community, state and federal level, as well as internationally. Community-based and all-volunteer rural fire services have been stretched to almost breaking point in the fires of 2019-20. The future may challenge them further. State emergency services, paramedics, state police forces and other first responder organisations face significant challenges recruiting and retaining people.

Now may be the time to consider implementing a voluntary but incentivised Australian Universal Scheme for National and Community Service (AUSNACS). Such a scheme could recruit from the more than 285,000 men and women who reach adulthood each year in Australia to participate as part of the three armed services, state or federal police, state emergency services, rural and country fire services, paramedic and more. Such a body would require coordination and endorsement from National Cabinet, but could generate significant benefits for the nation as it braces for greater challenges into the future.

Participants could be rewarded with student loan concessions or business start-up assistance as well as receiving trainee salary along the way. Such a scheme could see the disparate parts of Australia's multi-cultural community bond closer together. RFS/CFS firefighters in their off-season could help with nation park rejuvenation or assisting farmers with fruit picking, amongst other character-building activities.

Notwithstanding the many applications for organised forces in non-warlike settings, the threat of war still looms large and armed forces will still be needed. As this paper was being written, the threat of nuclear war appeared greater than it had for decades. Yet figuring out how wars end has continued to require the exercise a wide range of capabilities beyond just military power. As Morgenthau points out, there are numerous elements of national power that are exercised in the extension of politics by other means.

In essence, what has worked modestly well for Australia in years past likely will not be up to the task of what the future may hold. As a nation, a fresh sober-minded but visionary approach is required to enable us all to face the spectrum of emerging challenges ranging from looming environmental catastrophe, governance challenges and great power contestation. Yet in thinking about how wars may end in the future, it will help to have an awareness of the pitfalls of approaches taken in the past as other wars have ended.

Note

1 This chapter draws on the author's work undertaken in researching and writing including on Afghanistan and Iraq in *Niche Wars* (ANU Press 2020), the Korean War in *In from the Cold* (ANU Press 2020), the overlap of great power contestation, environmental and governance challenges in *A Geostrategic SWOT Analysis for Australia* (SDSC, ANU, 2019), Australia's involvement in Timor-Leste in *East Timor Intervention* (MUP 2015), Australia's post-Cold War involvement in operations near and far, including Bougainville and Solomon islands in *The Australian Army from Whitlam to Howard* (CUP, 2014), Australia's wartime experience from 1941–1945 in *Strategic Cousins* (MQUP, 2006) and its involvement in the Vietnam War in *Signals Swift and Sure* (1998).

References

Blaxland, J., 2002. 'Information-Era Manoeuvre: The Australian-Led Mission to East Timor,' *Journal of Information Warfare*, vol. 1, no. 3, 94–106.

Blaxland, J., 2019. *A Geostrategic SWOT Analysis for Australia,* Centre of Gravity Series No. 49 Strategic and Defence Studies Centre, ANU, Canberra.

Blaxland, J., Fielding, M. and Gellerfy, T. (eds), 2020. *Niche Wars: Australia in Afghanistan and Iraq, 2001-2014*. ANU Press, Canberra.

Breen, R., 1998. *A Little Bit of Hope: Australian Force Somalia*. Allen & Unwin, Sydney.

Breen, R., 2000. *Mission Accomplished: the Australian Defence force participation in the International Forces East Timor (INTERFET)*. Allen & Unwin, Sydney.

Breen, R. and Bou, J., 2016. *The good neighbour: Australian peace support operations in the Pacific Islands, 1980-2006*. Cambridge University Press, Melbourne.
Connery, D., 2010. *Crisis Policymaking: Australia and the East Timor Crisis of 1999*. ANU Press, Canberra.
Edwards, P., 1992. *Crises and Commitments: the politics and diplomacy of Australia's involvement in Southeast Asian conflicts, 1945-1965*. Allen & Unwin, Sydney.
Fukuyama, F., 1989. 'The End of history and the last man', *The National Interest*, no. 16, Summer, pp. 3–18.
Horner, D., 2011. *Australia and the 'new world order': from peacekeeping to peace enforcement, 1988-1991*. Cambridge University Press, Melbourne.
Horner, D. and Connor, J., 2014. *The good international citizen: Australian peacekeeping in Asia, Africa and Europe, 1991-1993*. Cambridge University Press, Melbourne.
Kingsbury, D., 2009. *The Price of Liberty*. Palgrave MacMillan, London.
Middleton, K., 2011. *An Unwinnable War*. Melbourne University Press, Melbourne.
Odgers, G., 1957. *Australia in the War of 1939–1945. Series 3, Vol 2, Air War Against Japan, 1943–45*. Australian War Memorial, Canberra.
Palazzo, A., 2001. *The Australian Army: A History of its Organisation*. Cambridge University Press, Melbourne.
Pelvin, R.H. and Straczek, J.H., undated. 'RAN in the Pacific War', Royal Australian Navy. RAN in the Pacific War | Royal Australian Navy
Ricks, T.E., 2007. *Fiasco: The American military Adventure in Iraq, 2003-2005*. Penguin, New York.
Ryan, A., 2000. *Primary Risks and Primary Responsibilities: Australian Defence Force Participation in the International Force East Timor*, Working Paper No 304. Land Warfare Studies Centre, Canberra.

11 How Major-Power Wars End

Hugh White

How a war ends depends, among other things, on who is fighting and what they are fighting over. This chapter explores how wars between great powers end, when they are fighting over fundamental questions of international order—what are often called hegemonic wars. This is a specific and rather narrow category because not all wars are fought between major powers, and not all major-power wars are fought for such high stakes. But it is an unusually important category because such wars are generally both the biggest, most destructive and the most consequential. How they end matters a lot.

Smaller wars between less powerful adversaries are often brought to an end by mediation or outside intervention, usually by more powerful third parties. But that doesn't happen when great powers fight, because there are no third parties strong enough to promote, let alone impose, a settlement. Fighting therefore only stops when one side or the other collapses, or when the costs and risks of continuing the war are so large that both sides accept a compromise. For that to happen the costs and risks must be very high, because great powers characteristically regard their place in the international system as extraordinarily important to their security, prosperity and identity—truly vital interests. It thus requires huge pressure to force them to compromise those interests. But in the nuclear age those pressures can be all too easy to apply.

The nuclear age was dominated for its first forty years by a strategic contest between two superpowers that were sufficiently equally matched both in power and resolve that neither side tried to contest the *status quo* between them until one of them unexpectedly collapsed. What followed was an interlude in which many people—especially in the West—believed that the era of great-power strategic rivalry had passed altogether, to be replaced by a unipolar global order in which America was, in effect, the sole great power. Today we know that was at best a passing phase, and probably always an illusion. America's claims to global leadership are being challenged in both Eastern Europe and East Asia by strong states that seek to assert themselves as great powers in their regions and as coequal players in a multilateral global order. For the first time, therefore, we find ourselves facing ambitious and powerful states seeking to contest the existing international order and armed

DOI: 10.4324/9781003317487-15

with nuclear weapons. The world is entering a new and more dangerous phase of the nuclear age when traditional competitive great-power politics is taking place between nuclear-armed powers. The risk of a major war between nuclear-armed great powers has quite suddenly risen sharply both in Europe and in Northeast Asia. And it is clear that in both cases the issues at stake go well beyond the immediate disputes, with profound implications for the whole basis of the post-Cold War international system. Countries around the world must now be asking themselves under what circumstances it might make sense to launch or join such a war, and a key question for them should be how that war might end.

We might best start by briefly recalling the backstory here—the way great power wars have evolved over the last few centuries.

Great Power Wars and International Order

The seventeenth century in Europe saw determined efforts, primarily by the leading Hapsburg and Bourbon dynasties, to establish hegemony in Europe. The last of those efforts—by Louis XIV of France, was finally defeated with the conclusion of the Wars of the Spanish Succession with the Peace of Utrecht in 1713-1715. Through the eighteenth century that followed, major powers often went to war with one another to achieve quite modest adjustments to the distribution of territory, power and influence. The outcomes of those wars did not affect, and were not intended to affect, the status of the combatants as 'great powers' in the European system. Even the most momentous of them—the Seven Years War—only resulted in a redistribution of territory and strength between the great powers, rather than a change in the membership of the Great Power club, let alone the establishment of an European hegemony. These wars were not carried all the way to anything like total victory or unconditional surrender. Instead they ended in complex compromise settlements. In other words, they were fought to diplomatic rather than military conclusions. In this era war really was, to slightly paraphrase Clausewitz, 'diplomacy by other means.'

The Napoleonic Wars of the early nineteenth century were different. As Paul Schroeder argues (Schroeder 1994) the old European order of the eighteenth century was coming apart even before this shock. That was one reason why Napoleonic France came closer than any state had ever come before in the modern era to achieving the kind of European hegemony that ambitious great powers had been seeking since the early sixteenth century. Defeating France required a series of wars of unprecedented scale, duration and cost, and those wars changed the way European states and their leaders thought about order and force. In Vienna in 1815 they created the Concert of Europe which, in various forms, moderated strategic rivalry between Europe's great powers and sustained a stable European order until the cataclysm of 1914. That does not mean that wars did not happen between the great powers in the nineteenth century. On the contrary, they were quite frequent, but the stakes were limited. The wars that broke out between European major powers between 1815 and 1914 were not fought over fundamental questions of international order. In particular, the status of the European great powers as great powers was not at stake. The same group of great powers that made peace in 1815 went to war in 1914.

The war that broke out in 1914 was the result of the pressures that had built up during the nineteenth century and made the old order built in 1815 unsustainable. A century of explosive economic growth, driven by the industrial revolution, had created massive shifts in the distribution of wealth and power between countries in Europe and beyond, and they were amplified by major political developments too. In Europe, the relative power of Germany and Russia had risen, while that of France, Austria-Hungary and Britain had fallen, as had the Ottoman Empire on Europe's periphery. And for the first time in modern history countries further afield—America and Japan—had become strong enough to affect the European-centric global order. One can see much of the strategic history of the twentieth century as the readjustment of the international system to these transformations in the distribution of wealth and power. This happened through two massive wars of quite unprecedented scale and ferocity, and the equally unprecedented four-decade period of intense and extraordinarily dangerous strategic rivalry that followed in the Cold War.

The pattern of great-power conflict in the twentieth century, and the way in which such conflicts ended, was profoundly affected by the way economic growth and technological innovation had changed the way wars could be fought. As the industrial revolution transformed the nature of warfare, the costs—in every way—of major-power war increased radically. Such costs were only to be contemplated for the highest stakes. The effect ran two ways, as we can see in the First World War. Major powers would only go to war with other major powers when the stakes involved the most fundamental questions. Once they were at war, the cost soon rose so high that any compromise that did not encompass fundamental changes in order was unacceptable in view of the sacrifices that had been made—witness the Western Allies' refusal to consider proposals for compromise peace with Germany. The result was that, in the twentieth century, major power wars have become rarer—indeed there were only two in the entire century—and both were fought over the most fundamental questions of international order. But once those wars started, they were fought to a decisive, indeed absolute, military conclusion, and resulted in the destruction of the political systems of the defeated countries. And the same could be said, in a different way of the third great strategic contest of the twentieth century, the Cold War.

Moreover, those three strategic contests—the First World War, the Second World War and the Cold War—all ended in a fundamental reordering of the international system, or at least so it seemed. At Versailles in 1919 Germany was deprived of its status as a great power, and an attempt was made to remake the whole nature of international relations by subordinating great-power politics and rivalry to the League of Nations. What actually happened was a renewal of the old contest after a 20-year interval. After 1945, Germany and Japan were deprived of great power status, and another attempt was made to transform the conduct of international affairs through the United Nations. What actually happened was the emergence of a unique bipolar strategic contest between the only two countries that had emerged from the Second World War with sufficient strength to count as first-rate powers. The contest between them ended with the collapse of the Soviet Union, the abandonment by Russia of the Soviet claim to great power status (or 'superpower' status in the language of the time) and the emergence—at least so it seemed—of a new, unipolar global strategic order led by the United States.

This was supposedly the 'end of history', but now we know that it was just a brief interlude. Instead of being uncontested, the US-led international order is now being challenged by rival great powers in two key regions. In Eastern Europe, Russia has sought to reclaim its status as a great power by asserting a sphere of influence over its 'near abroad' and contesting the post-Cold War redrawing of its Western borders. In East Asia, China now seeks to reclaim what it sees as its traditional role as the region's preeminent great power, and to push America out. These are seen, rightly, as the most important challenges to US leadership globally since 1989, and in some ways since 1945. How they are resolved, whether peacefully or by war, and if by war then how those wars end, will do a great deal to determine the shape of the international system in the decades ahead.

The Contest in Europe

Russia's invasion of Ukraine is one of the gravest challenges to international order since 1945. It is about a lot more than Ukraine, of course. Russia seeks to reassert its place as a great power and secure a sphere of influence over its 'near abroad' from which the United States and the European Union are excluded. These ambitions directly challenge America's post-Cold War vision of a unipolar global order. That vision sees America exercising preponderant influence in every part of the world, and thus rejects the idea that any other power should aspire to an exclusive sphere of influence. It sees US leadership of NATO, and NATO's strategic preeminence throughout Europe, as the foundation of the European strategic order. Russia's ambitions also challenge the European vision of Europe's future, in which the European Union sets the framework for economic, political and social development throughout the continent. The stakes are therefore very high. They have been raised further by the way in which Vladimir Putin has sought to achieve his ambitions. The blatant use of military power to seize Crimea and peel off parts of the Donbas in recent years was bad enough. The attempt in February 2022 to invade and occupy the whole of Ukraine and erase it as a sovereign state was much worse. And the brutality of Russia's military operations since the invasion, especially what seems to have been the deliberate and systematic targeting of civilians, has deepened Russia's offenses still further.

At the time this went to press, the war in Ukraine was not a shooting war between Great Powers, but it has certainly taken on many of the features of a Cold War between major powers, with severe economic sanctions and deep ruptures in almost all forms of normal interaction. What is more, the active and overt support provided by America and its allies through the provision of weapons and intelligence to Ukrainian forces has brought them closer to direct involvement in a war with Russia than we ever saw in the Cold War. There are several ways the crisis can go from here. One possibility is that Russia might suffer and accept a humiliating defeat. A second is a stalemate in which something like the present situation becomes a new de facto status quo. But a third possibility is that the crisis could escalate into a major power war between America and its NATO allies and Russia.

The potential for escalation to war between America and Russia is clearly very real. The initiative for such an escalation would most likely come from Russia. Its aim might be to isolate Ukraine and cut it off from Western support by convincing Washington to turn off the taps. Alternatively, its aims might be broader—to counter NATO's expansion and undermine the credibility of America's strategic position in Europe by testing America's and its NATO allies' willingness to fight a major war with Russia. President Biden's clear refusal to risk what he has called 'World War Three' by sending forces to fight in Ukraine might lead the Kremlin to assume that Biden would fail that test by backing down even if Russia attacks a NATO ally. Moreover, the willingness of many European NATO members to go to war with Russia over, say, one of the Baltic States is not to be taken for granted, and America would be much less likely to fight if the Europeans wouldn't. The temptation for Putin could be quite strong. But if he takes that gamble and his assumptions prove wrong, and America does fight, then we'd see the first serious war between major powers since 1945, and the first ever between nuclear-armed powers.

How would that war end? The surprisingly poor performance of Russian forces in Ukraine might encourage hopes that America and its NATO allies could decisively defeat Russia in conventional (non-nuclear) operations. But that is not to be taken for granted. Russia would have clear geographic advantages from fighting on its own borders. Its forces may learn swiftly from their recent failures and prove more formidable as time goes on, as they have often done in the past. But even if this does not happen and US and NATO forces scored significant conventional victories, it is hard to see how they could go on to deliver the allies' wider strategic objectives of compelling Moscow to abandon its great power ambitions and accept a subordinate place in the US-NATO-EU vision of European order. But neither is Russia, even if its forces do revive, likely to be able to inflict conventional defeat on America and its NATO allies sufficiently decisive to compel them to abandon their resistance to Russia's ambitions. The fact is that neither side is likely to be able to achieve their strategic objectives by conventional operations alone. That means both sides must either settle for a compromise or look to escalate beyond conventional warfare by using nuclear weapons.

That possibility is clearly encompassed by the nuclear strategies of both America and Russia. The Biden Administration has retained the 'First Use' policy that has been fundamental to US nuclear strategy since the early days of the Cold War America. Under that policy, America has long declared its willingness to initiate the use of nuclear weapons to defend its most vital strategic interests if its conventional forces fail to do so. Many would construe a failure to defeat Russia in a war over Ukraine as warranting such a step. But Moscow too has developed a doctrine of 'first use', and it is more likely than Washington to seriously contemplate that option. If Russia faces defeat in a conventional war, it is clearly possible that Moscow would look to its nuclear options.

Since the Cold War, Moscow has devoted a lot of attention to the way it might use nuclear forces to offset the loss of the clear superiority in conventional forces that it enjoyed during the Cold War. It has developed a concept called 'escalate to de-escalate'. The idea is simple enough: if Moscow believes that Russia's most vital interests are threatened and cannot be defended by its conventional forces, it will signal its willingness to fight a nuclear war by launching one or two nuclear attacks which, it is assumed, will convince the adversary to back off and abandon the fight (Schneider 2017: 368).

This idea has tended to be dismissed by Western strategists as fanciful or absurd because it would only make sense for Moscow to use nuclear weapons this way if it was very sure that America would indeed back off, rather than respond by using nuclear weapons itself. US strategists have tended to assume that Moscow could never be sure of that America would back down rather than fight a nuclear war, or at least not sure enough to justify the risk that instead of de-escalating a crisis, the use of nuclear weapons would instead lead to a swift escalation to a full-scale nuclear exchange (Episkopos 2021).

But US policymakers may well be wrong about this, because they may assume that Russian policymakers are as convinced as they are themselves of America's resolve to preserve US strategic preeminence in the post-Cold War order. They assume that US nuclear resolve is as unshakable today as it was in the Cold War, when Moscow could have been in no doubt that America would respond in kind to any Soviet nuclear attack. But it is clear that the US strategic community has not thought very deeply about these questions since the end of the Cold War and has thus failed to see how different the relative stakes, and hence the relative resolve of the two sides, are in today's confrontation from the way things were in the Cold War (Nichols 2022).

Back then the Soviet Union did seem poised to dominate the whole of Western Europe, and thereby pose a clear and present threat to America itself. The stakes for America in containing the Soviets in Europe were therefore truly vital, and even existential. Today Russia might be, at most, capable of dominating the states closest to Russia's own borders. It has no chance of dominating Europe's heartland, especially a united and powerful Germany. Russia's ambitions today thus pose no direct or vital threat to America, the way the Soviets did in the Cold War. That means it is much less credible—at least to Moscow's policymakers—that America would be willing to fight a nuclear war to contain those ambitions. And that makes their 'escalate to de-escalate' gambit much more plausible.

This is therefore a possibility that must be taken very seriously. There are of course two primary alternatives for what happens if it materializes. Either Moscow's gamble plays off, and America does not retaliate with nuclear attacks of its own, or Moscow's gamble fails, and the two sides find themselves in a nuclear war. There is no space here to explore the many ways in which the catastrophic scenario of nuclear war might play out, but we can identify two broad possibilities. One is that the nuclear exchange is quite rapidly brought to an end by agreement between the two sides. The other is that it escalates into a full-scale nuclear holocaust of the kind that loomed so threateningly during the Cold War. There is not much point in speculating about where that second possibility would lead, because it takes us far beyond the world of rational policy or strategy. But we can and should consider the three other scenarios—a conventional stalemate, a Russian nuclear strike to which America chooses not to respond and a limited nuclear exchange between America and Russia—would play out. How would such wars end?

What is striking about all three of these scenarios is that none of them present the kind of war-termination situations that have characterized full-scale major power wars since the dawn of the nineteenth century. There is simply no realistic prospect that either side could achieve the kind of complete victory, including the destruction of the losing side's system of government, that brought an end to the Napoleonic Wars, the First World War, the Second World War and the Cold War. Even if, as happened in Russia in 1917, the stresses of war bring about the collapse of the existing regime in Russia, any successor regime must be expected to maintain broadly similar objectives as their predecessors, as the Bolsheviks did after they recovered from the humiliation of Brest-Litovsk.

This means that the war could only be terminated by the achievement of a negotiated settlement, essentially between Washington and Moscow, which must be acceptable to both sides and would therefore require both sides to make major concessions to the other. We would thus find ourselves back in the world of eighteenth-century diplomacy, when wars were brought to an end by complex diplomatic compromises and painful concessions. In Europe today that means America and its NATO allies will find themselves accepting that Russia will achieve some kind of sphere of influence over its near neighbours, and that Russia will not accept US primacy at the global level, but will claim an equal place—in status if not in power—in a multipolar global order. The stark reality is that these outcomes, though deeply unwelcome, especially in Washington, are preferable to the costs and risks of war. And it means that the US-led European order of the post-Cold-War era has become unsustainable.

The Contest in Asia

The contest between America and China in East Asia is more serious and more dangerous than the contest in Eastern Europe. It is more serious, because China is far stronger than Russia, and its ambitions are bigger and more consequential. Russia does not challenge the entire European order but seeks only to contain its extent. China on the other hand seeks to overturn the entire strategic order in East Asia and the Western Pacific by replacing American hegemony with its own. Further, the outcome of this regional contest will have larger consequences for the global order because East Asia is such a large and dynamic region. Russia, if it succeeds in its ambitions, will still be a lot weaker and less influential internationally than America. If China wins the contest for East Asia, it will challenge and indeed overtake America as the strongest and most influential country in the world. The stakes for both sides are therefore very high. That makes the contest in East Asia more dangerous because it is more likely that America will decide that the stakes there are so high that they would compel America to go to war to resist China's ambitions. The dangers are further increased by the tendency of Washington to underestimate China as an adversary, and especially as a nuclear adversary, which will make it more willing to risk a war with Beijing.

War between America and China is not inevitable, but it is a clear possibility. That is because the contest between them will hinge on their willingness to defend or assert their opposing positions by going to war. Neither side wants war, but each hopes to use the threat of war to convince the other to back off. America hopes that its willingness to defend its position as the leading power in East Asia by force will deter China from pressing its challenge. China hopes that its willingness to fight to assert its regional leadership will convince America to concede. One side may get what they hope for, if in a crisis or series of crises the other side backs off, effectively conceding the claims of the other to superior power and resolve. The more confident each side is that the other will back off, and the more impatient they are to bring the contest to a conclusion, the more inclined they will be to provoke a crisis to test the other's resolve. On both these grounds China is more likely to take the initiative than America. But whichever side takes the plunge, it is clearly possible that when a crisis comes the other side does not back off, and war follows.

Today the most likely focus of such a crisis is Taiwan, and the discussion that follows will focus on that scenario. But it could just as easily be sparked by something else—a clash in the South China Sea or East China Sea for example. Whatever sparks the crisis, the key issue at stake will be the wider question of which of these two great powers will dominate East Asia. A US-China war over Taiwan, for example, would not really be fought over Taiwan, any more than the First World War was fought over the status of Serbs in the Austro-Hungarian Empire.

What kind of war would it be? Many in the West imagine that we have entered an era of grey zone conflict, in which major powers will be willing to limit their engagements to the realm of cyber-war and to the use of non-military and paramilitary forces. This is wishful thinking. Both America and China have massive conventional armed forces specifically prepared for such a conflict, and when the stakes are so high they would both be impelled to use them. We must expect that, unless one side backs off and concedes defeat, any substantial clash between them would quickly draw in major conventional forces and swiftly escalate towards a full-scale war. Then the chances of defusing the crisis by both sides backing off would be low, as each side would hope to score a decisive diplomatic and strategic victory by forcing the other side to do so.

The world would then be in unchartered waters. This would be the first serious war ever between nuclear-armed powers. It would be the first serious war between major powers since 1945. And because America has no capacity to fight a continental war on the mainland of Asia, it would be a maritime war, and the first significant maritime war since 1945. Much has changed in the technology of maritime (air and naval) warfare since then, as both ships and aircraft have become both easier to detect and easier to destroy. The net effect of these changes had been to shift the balance of advantage strongly towards the strategic defensive and against the strategic offensive. The side that seeks to project power by sea and air has a much tougher task than the side that seeks to stop them (White 2015). In a US-China war this advantage favours the Chinese, because they can fight from home bases while America must project power by air and sea over long distances. Moreover, Beijing has invested massively, and it seems very effectively, in developing and building air and naval forces to exploit its inherent advantages at America's expense.

This has revolutionized the military balance in East Asia, with huge implications for the trajectory of a US-China war there. Ever since it defeated Japan in 1945 America has enjoyed unchallenged preponderance in air and naval force in the Western Pacific. Today that is no longer true. Washington can no longer expect that a war with China would lead swiftly and easily to a decisive US victory. Instead, it is most likely that a war over Taiwan would lead, after a week or two, to a costly stalemate. Both sides would have inflicted heavy losses on the other, with many ships, aircraft and bases destroyed, and many lives lost. But neither side would have inflicted enough damage to compel its adversary to withdraw from the fight. Nor would there be any serious prospect that further conventional operations would make any difference to that.

How then could the war be brought to a conclusion? There are several possibilities. One is that one side or the other accepts defeat and withdraws. That seems unlikely. Both sides would have suffered heavy losses, probably including attacks on their sovereign territory. America would almost certainly have launched a sustained strike campaign against Chinese bases on the mainland, and China in turn would have hit the vital US base on Guam and perhaps even launched attacks on Hawaii. Added to the immense stakes which drove the two sides to war in the first place, these losses would make it hard for either side to abandon the fight at this stage. But if either side does, it is more likely to be America.

It also looks unlikely that the two sides could simply allow a stalemate to fossilize into a new status quo. Both would then have to look for options to break the stalemate and prevail. American options to do that look very limited. There is no reason to expect that even a major and sustained conventional bombing campaign would do enough damage to drive Beijing to surrender. Nor is an economic blockade likely to work, and the chances of an internal political insurrection in China favouring America are remote indeed. China's options are no better. So for both sides, the only way to break the stalemate would appear to be to look to their nuclear options. There is a parallel here with the contest in Europe. It seems that in neither theatre does either side have conventional forces capable of forcing their adversary to concede when the issues at stake go to such fundamental issues of international order. Hence in both theatres the possibility of escalation to nuclear war soon becomes a vital consideration.

There is a difference however in the calculus of nuclear escalation in the two theatres. In Europe, as we have seen, it is Russia that seems most likely to look at nuclear options first, primarily because, while US and Russian nuclear capabilities are broadly equal, Russia's stake in the outcome is clearly greater than America's. In East Asia, it is likely that America would look to the nuclear options before China, because its nuclear forces are so much bigger and more capable than China's, and it is less clear that US resolve is inferior to China's. The difference on the balance of resolve in the two theatres is shown by the fact that, in Europe, President Biden has explicitly and emphatically ruled out any US combat intervention in Ukraine, whereas he has very deliberately not done so in regard to a Taiwan contingency, and has indeed at times gone beyond the official US policy of ambiguity on this point to explicitly say that America would fight for Taiwan. Below is discussed how these twin issues of the balance of nuclear forces and the balance of resolve between America and China might actually play out as an escalating crisis approaches the nuclear threshold.

As has been noted, the possibility that America would use nuclear weapons first in a war with China is explicitly encompassed by its longstanding 'First Use' policy. But it is unlikely that US policymakers would envisage actually using a nuclear weapon, at least not at initially. They would more likely plan to simply threaten nuclear attack, expecting that this would be enough to make Beijing back off. Their confidence that Beijing would not be willing to risk allowing the war to cross the nuclear threshold is based on the fact that America's huge nuclear arsenal means that it could inflict far more devastation on China than China could inflict on America. It is no doubt true that China would not contemplate fighting a nuclear war with America, and that is confirmed by the fact that it has never sought to build a nuclear arsenal that approaches America's.

But that does not mean that US threats of nuclear attack would necessarily be enough to make Beijing concede defeat in a US-China war, because Beijing may not believe that Washington's threats are credible. China's nuclear forces may be smaller and less capable than America's, but they are big enough to pose a serious threat to the continental United States. At a conservative estimate, China could deliver major nuclear attacks on half a dozen US cities and kill hundreds of thousands of Americans. Beijing may well believe that the likelihood of such a retaliatory nuclear strike would deter Washington from carrying out a nuclear threat. In fact, they almost certainly do believe that because ever since China became a nuclear power in the 1960s this has been the primary function of its nuclear forces. Its posture of 'minimum deterrence' has been designed specifically for this situation. It aims to counter the kind of nuclear blackmail that Washington would rely on to break a conventional military stalemate by undermining the credibility of a US nuclear threat by threatening nuclear retaliation.

US strategists have long sought to counter China's minimum deterrent by looking for ways to neutralize a Chinese retaliatory strike by preemptively destroying Chinese missiles before they are launched and shooting them down in flight. But the Chinese have continued to expand and develop their nuclear arsenal in response, and it is only prudent to assume that they believe they have succeeded in preserving the capacity to inflict enough damage on America to deter a US nuclear strike on China, because if they didn't believe that they would simply have built up their nuclear arsenal faster which they have plenty of capacity to do. And, of course, it is what the Chinese believe that matters here, because if they do not believe the US will launch a nuclear attack, they will not back off in the face of US threats.

This is not, however, the end of the story. Washington might be able to neutralize China's threat of retaliation and thus restore the credibility of its nuclear threats by threatening to launch counter-retaliatory nuclear strikes against China in response to any Chinese retaliatory strike on America. This is where things get complicated. How sure could US decision-makers be that Beijing would respond to this counter-threat? There are two ways they might try to neutralize it. They might try to deter US counterretaliation by holding back a proportion of their forces from their first retaliatory strike with which to threaten a second, counter-counter retaliatory, strike in order to deter US counter-retaliation. Or they could simply decide to accept America's counter-retaliation. That is not as unlikely as might at first appear. It must be remembered that throughout the Cold War America's nuclear strategy depended on their evident willingness to accept a massive Soviet nuclear attack on the continental United States rather than concede control of Western Europe to the Soviets. There is no reason for US decision-makers to assume that China would be any less willing to accept US nuclear attack on China rather than concede defeat at America's hands in a war in East Asia.

Clearly there are limits to how far it is useful to pursue speculation about the endless permutations of threat and counter-threat which would come into play in this kind of nuclear stand-off between China and America. But two things are clear from the limited analysis offered here. The first is that America's inability to achieve a decisive victory over China in conventional air and naval operations in the Western Pacific means that it would have no clear path to achieving its strategic objectives in a war with China other than by threatening nuclear attack. The other is that such threats are unlikely to be effective, because they are unlikely to be credible to China. They would only be credible to Beijing if the Chinese believe that America would be willing to accept a Chinese retaliatory strike and would thus not be deterred from launching a nuclear attack by China's threat of retaliation. Again, there are lessons from the Cold War here, because in that struggle, as we have noted, America was able to convince the Kremlin that it was willing to accept a full-scale Soviet nuclear strike rather than allow Moscow to dominate Europe. Indeed, that was the essential condition for the effectiveness of America's First Use policy in countering the Soviet's superior conventional forces.

The question therefore arises, whether US policymakers could in the same way convince the Chinese leadership that they are willing to accept a Chinese nuclear strike on the continental United States rather than allow China to dominate East Asia. If the stakes for America in East Asia today are as high as they were, or were seen to be, in Europe in the decades of the Cold War, then the answer would be 'yes'. In that case a US threat to use nuclear weapons first would offer a clear and credible path to victory in an East Asian war. But if not, then it is hard to see how America can break a stalemate and prevail over China. Which is it? The key to America's determination to contain the Soviet Union in the Cold War was the conviction that America's own direct security was at stake. Washington feared that if the Soviets could dominate Western Europe, then they could dominate the whole of Eurasia, which would give them such an overwhelming preponderance of power that they could directly threaten the United States itself. This was not an unreasonable fear, at least in the earlier decades of the Cold War. It was, at any rate, accepted not just by leaders and policymakers in Washington, but by the great majority of the American people, who clearly understood the nature and scale of the risks they were accepting to confront Soviet ambitions.

It is far from clear that the same can be said of China today. As we have seen, China's ambitions to dominate East Asia directly challenge both

America's long-standing leadership in the region, and its claims to global primacy. But China does not pose the kind of potential danger to America's security that the Soviets could have posed in the Cold War, because China for all its wealth and power does not have the prospect of dominating Eurasia as the Soviets might once have done. Today Eurasia contains too many other powerful states to make that a serious prospect. This means that even if China can take America's place as the leading power in East Asia, it will not be able to accumulate the power to threaten America itself. And that means America does not have the same compelling need to contain China's ambition as it did to contain the Soviets—not compelling enough to impel it to accept a nuclear attack from China. That perhaps explains why, as political leaders in Washington have talked up the need to confront China strategically, they have not been willing to make clear the nuclear costs and risks involved. All this makes it unlikely that the Chinese leadership can be convinced that US resolve in this 'new Cold War' is as strong as it was in the last one. And that makes it hard to see how America can use nuclear threats to win a war with China, and hence how it can win a war with China at all.

So much for America's nuclear options. But what of China's? For a long time it was assumed that China would never initiate nuclear war. Not only does its declared policy of 'No First Use' preclude such a step, but the forces it developed to implement its posture of 'minimum deterrence' was ill-suited to nuclear warfighting. But that has started to shift in recent decades, and while Beijing still declares that it will not use nuclear weapons first, the shifting balance of capability and resolve in the post-Cold War world mean that there are now clear scenarios in which they might be tempted to do so. Just as America's reduced strategic stake in Europe makes it credible that Moscow might try to implement its 'escalate to de-escalate' gambit, Beijing might come to believe that it could use nuclear weapons to break a conventional stalemate in East Asia. China's leaders might consider, for example, that they could launch a nuclear attack on a US aircraft carrier or even against the US base on Guam, and then deter any US nuclear retaliation by threatening counter-retaliation against the United States itself. The same complex issues would then arise as we saw in considering America's option for nuclear first use, concerning the credibility of each sides' threats of retaliation and counter-retaliation. But in this case China would have a clear advantage that America lacks: its threats are more credible because its stake in the outcome is higher. In the end this boils down to a simple question of geography. What happens in East Asia matters more to China than it does to America because it is on China's doorstep—just as what happens in the Western Hemisphere matters more to America. Washington would have the advantage in a confrontation on America's doorstep—as was seen in the Cuban missile crisis. In a war over Taiwan that advantage would lie with China. That means the temptation for Beijing to use a limited nuclear strike to warn Washington of the potential for nuclear escalation and scare it into conceding defeat would be very real.

Several important conclusions about how a US-China war would end can be drawn from this brief analysis of what would be a very complex situation. The first is that, in the absence of a clear advantage in conventional military forces and operations on either side, the threat of nuclear weapons is very likely to loom as the key factor in deciding the outcome. The second is that, while the incentives to avoid nuclear war would remain strong, the chances that such threats would materialize into the actual use of nuclear weapons is quite high. On some plausible assumptions about US resolve, the Chinese side might well conclude that limited nuclear strikes could provide a path to victory. And under the extraordinary stresses of a crisis, either or both sides could act irrationally and launch nuclear attacks even when that offers no credible hope of furthering their objectives. And once both sides have used nuclear weapons there is no reason to hope that a full-scale nuclear holocaust could be avoided. China would suffer far more, but the consequences for America would still be unimaginably devastating.

This leads us to some broad conclusions about how a US-China in East Asia war might end. The chances of a decisive conventional victory by either side are very slight. Likewise, it is hard to imagine that the two sides would find a way to allow the stalemate to be perpetuated as a new status quo. So, unless one side or the other—most probably America—backs off at this point and concedes victory, and primacy in East Asia, to China, then the conflict will be resolved by the use or threat of nuclear weapons. The key conclusion therefore is that avoiding a US-China war must be a high priority—a much higher priority than it has appeared to be in recent years. Further, that is a responsibility that falls on both sides. As E H Carr noted many years ago, defenders of the *status quo* often assume that it is their adversary's fault if war breaks out (Carr 1939: 191). But defending the status quo no more inherently justifies resorting to force than challenging it does. America and its allies are equally responsible with the Chinese for steering away from the abyss of major-power war, and common sense suggests that they should be willing to make substantial compromises in order to do so.

Conclusion

The key conclusion of this brief study of how major power wars might end in the future is that they are likely to end in a nuclear exchange, and quite possibly in an uncontrolled nuclear holocaust. This should not come as any surprise to anyone who recalls the Cold War, but it seems to need to be learned anew by a later generation. The hope of a post-Cold War order free of major power rivalry having proved illusory, we must now confront anew the realities of major power conflict in the nuclear age. And we must recognize that as the US-led global order falls apart, the resulting major power rivalries lack the symmetries of power and resolve which, we can see in retrospect, kept the nuclear peace between the superpower for forty years until 1989. As America seeks to keep its place as the globally dominant power against the regional ambitions of Russia and China, the risk of nuclear war seems to have grown swiftly, and in a way that has taken leaders, policymakers and analysts by surprise. It is time we woke up to the perils we face, and the urgent need to avoid them by managing the emergence of a new multipolar global order that better reflects the real distribution of power and resolve in the world today.

References

Carr, E.H., 1939, 1981. *The twenty years' crisis 1919-1939*. Macmillan, London.

Episkopos, M., 2021. 'Russia's crazy nuclear war strategy: escalation to de-escalate?' *The National Interest*, March 19. https://nationalinterest.org/blog/buzz/russia%E2%80%99s-crazy-nuclear-war-strategy-escalationto-de-escalate-180680 [accessed 12 June 2022].

Nichols, T., 2022. 'We have no nuclear strategy,' *The Atlantic Monthly*, June 1. https://www.theatlantic.com/magazine/archive/2022/07/us-nuclear-strategy-cold-war-russia/638441/ [accessed 12 June 2022].

Schneider, M.B., 2017. 'Escalate to de-escalate,' *Naval Institute Proceedings* February, Vol 143/2/1. https://www.usni.org/magazines/proceedings/2017/february/escalate-de-escalate [accessed 12 June 2022].

Schroeder, P., 1994. *The Transformation of European Politics 1763-1848*. Oxford University Press, Oxford.

White, H., 2015. 'The Maritime Balance in Asia in the Asia Century', in G. Till (ed.), *The changing maritime scene in Asia—rising tensions and future strategic stability*. Palgrave, London. pp 9–21.

Section 4
Ways Forward

Introduction

The last section of this book looks at some ways forward, and some pitfalls that can occur along the way. William Zartman begins by putting a case for negotiated settlements to wars, noting that even where there is a preponderance of military capacity, most wars end by negotiation. He poses a series of questions, or tests, which are fundamental to the negotiation process, including the point at which respective parties are ready to negotiate, who speaks on behalf of those parties (particularly for non-state actors), how hard they can bargain and at what point do they achieve sufficient to be satisfied that an end is in their interests.

Damien Kingsbury follows with sometimes overlapping set of issues in negotiations, including the preparedness of the parties to be at the table, the role of wider political change in negotiations, readiness and capacity to negotiate and the point at which wars produce costs without the prospect of gains for either side, as well as the potential roles of third parties in bringing about a resolution. This section concludes with Adam Day and Charlie Hunt considering the role of the UN in peace-keeping operations, what seems to be the increasing intractability of many conflicts and their propensity for relapse. How UN peacekeeping operations are conducted, they suggest, has lessons for their longer-term success.

DOI: 10.4324/9781003317487-16

12 Negotiations To End All Wars

William Zartman

Most wars end by negotiation. Parties come to the conclusion that the war is costing more than it is winning or likely to win, and so decide to negotiate, a Ripeness Question. But negotiation to end wars is a three-act drama. In addition to the negotiations between the sides, each side has its own negotiations to handle the triple question: who speaks for each side and is able to deliver its compliance (the Valid Spokesman Question)? Is the appropriate tactic for maximizing benefits and reducing costs one of hard bargaining or accommodation (the Toughness Dilemma), and what is the acceptable outcome? (the Enough Question), all related but separate questions. Thus the matter of war termination involves the major questions of negotiation analysis.

The *Ripeness Question* is the broadest consideration in coming to a war's end. Parties of a war make a calculation, explicit or implicit, of cost and gain of the enterprise. Current scholarship seems to focus on the question of why weaker parties go to war, but this is just a minor segment of the larger cost-and-gain question. The tricky thing about this calculation, of course, is that it has to do with projections, uncertainties about uncertainties, not with any figures verifiable until well after the fact. But there are even greater complexities about this question in wartime, because it must be answered by both sides when each party's calculation depends on the other's and one party may feel that it is doing better in the conflict than the other and than it really is.

Yet negotiation, even in wartime, is occasioned by a common feeling that the conflict is in a Mutually Hurting Stalemate (MHS) in which unilateral pursuit of a solution is unpromising and painful (costly) and the possibility of a shared Way Out is a shared perception. Such projections are not only difficult to calculate to a point where a decision is possible but they are even more difficult to communicate to the other party. Indeed, communicating the feel of being in a hurting stalemate that is not mutual is likely to be seized by the opponent as a sign of weakness that will be seized by the opponent to break the stalemate. In the absence of a direct announcement, it is likely that each side will rely on intelligence findings and then on secondary, objective signs such as losses in money, lives, and public support

DOI: 10.4324/9781003317487-17

that may then be internalized by the party as the basis for a sense of an MHS. But above all, hurt, stalemate and ways out are subjective perceptions; the only objective reality is the action taken on the basis of these perceptions.

Once the negotiations are in course, the opening of the endgame could also be analyzed as a ripe moment. If the parties are blocked in the course of their negotiation and find themselves again in a stalemate, they need to revive the conditions of mutual pain. The push to negotiations may need to be renewed during the process, causing a reversal in the atmosphere of the negotiations that could disrupt positive movement, but this may be necessary if the pull of a sense of a way out and its development into a mutually enticing opportunity is to be maintained. In this sense, ripeness for closure is not an externally defined contextual situation like MHS but an internally created stage built up by the process itself: 'if we fail now, we will all be worse off.'

Discussions of ripeness tend to focus on the element of hurt and losses, with less attention paid to estimates of gain and future goals. Such consideration involves questions such as 'What are we in this for?' 'If we can only hang on now, things will be better in the future,' 'No gain without pain,' and 'We can't give up now, the gain is worth any cost.' These are among the most difficult of the uncertainties because they involve not only costs and goals now but the uncertain costs of uncertain attainability in the future. Even if a stalemate now is admitted, in a situation of increasing cost, the goal can make all pain bearable. The result is that the MHS can be clearly conceived as a time when the perceived-gain line dips below the rising perceived-cost line, but the heuristic image of lines hides a multitude of woolly uncertainties. Furthermore, these lines are clothed with a number of substantive and procedural wrappings of ends and means that further complicate the clear image of a crossing point.

Finding a ripe moment shifts the negotiation arena to the insides of each party, beginning with the question of the *Valid Spokesman*, an additional element not always mentioned in the concept of ripeness. These 'push and pull' perceptions and decisions will be taking place during continued and even renewed campaigns to support the war effort, hence raising the bar for the acceptance of a sense of stalemate and of hurt; it becomes very difficult for a party to reverse course when its course has been so assiduously nurtured up to this point. Leadership in the war often becomes a matter of personal as well as political commitment. Thus, the MHS has often been accompanied by a change in leadership that makes a change in course possible, either within the established decision-making apparatus or by even a revolt. The process of finding a valid spokesman may involve engagement of one party in the internal politics of the opponent, facing the danger that the eventual choice may run the risk of being compromised by external interference.

The process requires each party to find spokesmen who are willing to admit an MHS and engage in negotiations but who are also able (valid) to bring their party along with it to a decision. Thus the Valid Spokesman Question is not just a matter of the choice of an individual but the need to assure a supportive base within his/her constituency. However, in fact, public and supporters' opinions of the war may well change before those of the incumbent leader. The putative spokesman may then pick up this change through an upcoming election or, eventually, through coup or even revolt.

The *Toughness Dilemma* is also a procedural question that shapes the decision to talk and runs along with any negotiation process; in fact, the Toughness Dilemma is the negotiating form of the more basic Tactical Question: talk or fight? Is it better to hang tough to improve the outcome but to risk the danger that there be no outcome at all, or to lean soft to improve the chances of an outcome but lessen the chances of getting a good share of it? The choice of tactics influences the other party's choice, and shifting approaches becomes confusing. Among the modes of negotiation, hardliners frequently adopt a zero-sum attitude, use a distributive style, and negotiate by concession, whereas softliners seek positive-sum solutions, pursue integrative bargaining, and negotiate by compensation (trade-offs). Significantly, the parties can also consider construction and reframing their goals so as to make them fit into a joint agreement without changing their substance. Each of these elements includes a range of behaviors and tactics involved in turning the decision to end the war into a joint agreement to do so.

The Question and then the Dilemma is a continual subject posed during the war and the negotiations and the values of either option change as the conflict and talks go on. The decision to talk is not irrevocable (only the decision to go to war is). The decision to talk tough or soft is variable and tactical, but too much variation is confusing to the opponent and hence to the progress of the negotiations. Parties get a reputation for a bargaining style, and the opponent is prepared to act in function of that expectation; unfortunately, both a correct expectation of hardness and an incorrect expectation of flexibility will push the party into hardness of its own.

The *Enough Question* is the substantive companion of the previous tactical questions. There is no clear notion of a bottom line or a red line in negotiation, despite its frequent reference. The job of negotiations, like the job of war, is to move the other party's notion of a bottom line, while defending one's own. 'This far and no farther' can be the prelude to tactical retreats and deals through compensation to improve one's total outcome. Thus a party does indeed have in mind a whole range of outcomes, from the lowest acceptable payoffs now to the broadest goals on the future horizon.

If there is anything that gives a solid referent for acceptable goals, it is the outcome that a party could achieve without negotiating, termed a 'Best Alternative to Negotiated Agreement' (BATNA). But parties often conduct negotiations and also warfare against each other's BATNAs if they cannot inflect positions directly; *faits accomplis* often change the game board and what was considered the best alternative attainable without negotiation no longer is. War accompanying or interrupting negotiations is an effort to enforce or change a party's BATNA or to lower the opponent's notion of its best alternative to negotiating. In addition, BATNAs can be considered the best proxy for power. If a party can get enough of what it wants without negotiating, it is in a stronger position than if it needs negotiation (and hence needs to pay more) to get what it wants.

Thus 'how much' involves the degree of the goal (in some sense) that a party is willing to settle for; it also involves the price. How much does one have to pay to end a war, by any combination of means—the cost of the gain—in the final settlement compared to the cost of continuing the war, the two sides of the Enough question. Does the party have enough to pay for peace, compared with what it is paying to wage war, and what does one get along with an end to hostilities? Violence is the coin or bargaining chip of one or both parties, which it seeks to trade in for a realization of its goals; most frequently it is the club of the rebels, as the government seeks to recover its monopoly over the legitimate use of force but give up its use against the rebels. The juncture of the two sides of Enough brings the consideration back to the crossing of the two lines of cost and gain, no longer at the beginning of the ripening process but toward the end as the parties bring their two answers together in an agreement that (ostensibly) ends the war.

These conceptual questions involve calculations that the parties make as they move to a consideration of continuing or ending a war. There is no telling where they will come out but there is an understanding of what they have to go through to get there that can explain why certain paths were taken or not, and in the end can explain how they came out when they did. But there is a final dimension, as indicated at the beginning. These calculations are not simply between the two or more parties; they are also going on within each party. Parties have sides. One of the concepts, the valid spokesmen, indicates that there are other spokesmen who are not valid and that validity is a contested value. There is some recent literature on fragmentation of parties in conflict but the subject is only indicated, by no means completed. The questions discussed above have their place in each side's internal debates, but there is no clear structure for the analysis of these local negotiations, despite good beginnings with boundary role conflicts and two-level games.

These basic concepts have been laid out rather sharply to get the ideas clearly across. But concepts are square and reality is not; its shape and boundaries are always muddy. As with any tools: it is worthwhile to try them out on reality to see if they can cut through the fog of war and the marsh of peace, and address the problem of getting from one to the other on the ground. Several cases will be evoked, briefly, not as full histories but to bring out the characteristics of the war-ending process, before returning to a review of the concepts in a dynamic context.

Wars these days are major power conflagrations as well intrastate conflicts against non-state armed groups (NSAGs) with interstate involvement. Big war has not yet provided an ending to analyze, but a ready collection of internal conflicts where the government is faced by NSAGs rebelling for either a change in the central government or secession defined by territory or population. Their goals can be programmatic or identity-based, grouped together under the label of self-determination in a situation where the aggrieved population feels that their needs can no longer be met in the incumbent authorities and the whole or part of the country must be put in their own hands.

Case Studies

Colombia

Colombia has been the site of deep social conflict in the post-World War II era. While the rivalry between two elite factions locked in La in 1946-58 was ended by a civil pact of alternance, it also excluded poor rural populations from the political system. Promoting their interests and then inspired by Marxist ideology, they formed the Revolutionary Armed Forces of Colombia (FARC) and other groups in 1964, launching a long and bloody guerrilla conflict in which the Campesino rebels soon controlled large portions of territory. Beginning in the 1980s, governments felt themselves in a painful stalemate, which was however not mutual: the FARC enjoyed its autonomous status and inflicted damage on government and populations. When it found a need for increased resources, it joined the drug tradein1982; while it became hooked economically, it never lost its ideological and programmatic calling. Attempts to end the war were made by the governments of Belisario Betancur, Virgilio Barco, Cesar Gaviria, Andres Pastrana and Alvaro Uribe.

The last presidency, with Juan Santos as its defense minister, exacted heavy damage on the FARC and at the same time produced an amnesty for the right-wing self-defense forces (AUC). The military situation of the FARC worsened, and at the same time the NSAG changed leadership with the death of its longstanding committed communist for an ideological successor and then for more flexible but still dedicated leader. As Uribe's successor, Santos felt an opportunity for the opening of talks; a MHS had arrived in 2010 and with it a shared sense of a way out, to be fleshed out in negotiations. The mutual feeling was made possible by a new spokesman for each side, supported by each side's leadership structure.

The evolving agreement was made by acceptance of two major FARC demands in a form acceptable to the government. Participation (a point offered as early as Barco) was offered with electoral participation as a party and some guaranteed seats and the agrarian question was subject to serious reforms. On the government side, a demand was for prosecution of criminal behavior during the insurrection and an end to violence (Barco's other term). Together the four points made a formula for agreement. Negotiation followed a driving mode with some incidents involving violence handled gently. There was much debate on both sides as to whether the concessions were too much, rather than just enough. There was much criticism from the opposition led by Uribe against agreeing to prosecution for criminal action by the government (or UAC) and from the FARC side against the same provision.

Once agreed upon, these debates took legal form. The agreement was rejected in a referendum; some changes were made to tighten the government position and the FARC was told, 'reject the changes and you get nothing, only return to war.' The agreement was then passed in the parliament, circumventing the referendum. In the following elections, a Uribe follower was elected president with a promise to revise; he did not revise, but implementation by both sides dragged, in part because of lack of resources for the reforms and in part because of local rebel resilience. A few FARC signatories denounced the agreement but received no major support. The war is over, but localized violence unrest continues. The dogging problem was the question of valid spokesman: was Santos representative of his whole country? He seized the ripe moment but was able to carry only half his population with him.

South West Africa

South West Africa was a League of Nations protectorate under South Africa, but the mandate was removed by the UN and recognition of the South West African Peoples Organization (SWAPO) was recognized as the sole legitimate spokesman for the population. The US under President Carter attempted a 'resolution before the conflict' while SWAPO was weak and inactive, even threatening to break relations with South Africa if it went ahead with elections with its favored party and without SWAPO—which it did but which the US did not. However, the war picked up with the assistance of the Cuban military in Angola (where SWAPO was mainly cantoned) and the South African Defense Force (SADF) pursed the enemy deep into Angola. South Africa said its troops were occasioned by the Cuban occupation of Angola, and Cuba claimed its troops were occasioned by South Africa's invasion; each side would welcome the other's withdrawal, under an Alphonse-Gaston act. Under Reagan, the US returned to mediate, under the slogan of 'constructive engagement,' meaning that pressure would be put on South Africa as well as Cuba/Angola (who would bring SWAPO along). In 1986, a battle at Cuito Carnavale ended in a drawn; body bags began coming back to South Africa and Cuba threatened hot pursuit into South Africa proper; at the same time Angola, in Moscow, negotiated an increase in Cuban troops, but let it be known that it was sick of them being in Angola (kleptocratic, nepotistic, and not communist enough). Finally in 1986, the MHS arrived, under six years of patient nudging by Assistant Secretary of State Chester Crocker.

The two sides, sensing a common belief in a way out, began to listen to the mediator. The bargaining was tough dueling every inch, as both leaders had to show their followers, before whom they had long argued a now reversed policy, that they were fighting hard every inch. Continued sporadic violence only underscored the MHS. The formula for the agreement in 1988 was simply simultaneous staged withdrawal of both sides' forces, which had reached the same level of 50,000 each, with the additional Cubans' arrival, and the independence of South West Africa as Namibia under SWAPO. But it was enough for both sides. Within slightly over a year of the agreement, South Africa, under new leadership, renounced, the apartheid system (it took Angola another thirty years to get a new leader and maybe longer for a new system).

Mindanao

The major island of the Philippines, Mindanao lies to the south and is populated mainly by Muslims; Spanish and then US occupation brought in their own people to colonize the formerly independent sultanate of Sulu. Muslim (Moro) consciousness arose after the war, leading to conflict over landownership, augmented by conflict over family or personal honor (*rido*), and broke out into rebellion under the Moro National Liberation Front (MNLF) in 1972 against repressive measures of the Ferdinand Marcos. Libya, as chair of the Organization of the Islamic Conference, stepped in as a mediator but beyond a ceasefire, negotiations dragged on, and the Moros broke into more radical factions. When Marcos was overthrown, the successor government of Corazon Aquino passed a constitution providing *inter alia* for an Autonomous Region of Muslim Mindanao (ARMM), which the break-away Moro Islamic Liberation Front (MILF) rejected as insufficient. As the conflict continued, Indonesia mediated an agreement with the government of Fidel Ramos in 1996 to implement the constitutional provision, which the MILF continued to reject with renewed violence, met by 'total war' by the following government of Joseph Estrada. The next government of Gloria Macapagal Arroyo, returned to negotiations with the MILF to produce an MOU on Ancestral Domains, which was overturned by the supreme court for being too close to secession; vicious conflict continued. Then, in 2010, Benigno Aquino began negotiations under a new Malaysian facilitator to produce a framework agreement in 2012 and a succession of framework annexes leading to a comprehensive agreement for a Bangsamoro (Moro Nation) in 2016.

An MHS in Mindanao was difficult to produce; it took successive collapses and renewed violence and was arguably aided by the frustrating collapse of successive agreements as much as by the revival of 'total war'—ultimately unsuccessful—by a hardline president who followed a more resilient predecessor, a seesaw on the Tactical Question produced by the Philippine practice of reversing the record of one's predecessor. The practice highlights the importance of the valid spokesman; the process was controlled as much by the step back by the hardliner as by the step forward by the successor, independent of the issues of the conflict. The ensuing process highlighted another important aspect of war-ending: the question of enough. The successive negotiations and mediations were a repeated search for an adequate formula that was liberal enough to satisfy the Moro nationalists but short of secession, prohibited by the constitution. In the end, President Aquino announced that the agreement had 'created a ... Bangsamoro,' unable to find an appropriate translation of the term (Hampson 2015).

Afghanistan

Afghanistan presents a control case. It is hard to know on which war to focus on. Although the nineteenth century was filled with two major wars with Britain (from which we learned nothing), the unending war of the twentieth century began with the overthrow of the monarchy in 1973 and the 'protective' intervention by the Soviets in 1978. Armed resistance by Islamist groups and tribal militias began immediately; the ideological gap between the two sides was unbridgeable, but the divide between the moderate and the Islamist mujahideen was almost as wide. Following a UN mandate of 1980, the UN Special Envoy and the Personal Representative of the Secretary-General (SE/PRSG) began six years of shuttle mediation in 1982. It is hard to pinpoint the arrival of a sense of an MHS, which was rather an on-and-off sense looking for the best balance of forces for negotiation. Uniquely, the hurting stalemate was above all political for the Soviets and their Afghan government as both came under rising international opprobrium. But in the early 1980s, Moscow concluded that Kabul, not strong enough to win, was not strong enough to negotiate, and so the two sides fell back on reinforcements and revived hostilities, raising the cost for Pakistan but also increasing friction within each side. As the sense of a costly stalemate grew and with it the sense of a shared way out toward the later 1980s, the mujahidin broke out of their anti-Soviet unity and split over control of the post-agreement government.

The search for terms of the agreement was sinuous since the ideological as well as interest differences among and within the parties were absolute. The UN mandate formula was nonaligned sovereignty, territorial integrity and the right to choose a course; prohibition of external intervention; foreign troop withdrawal; and refugee return. Pakistan's proposed formula was Soviet withdrawal; internationally guaranteed non-interference; and refugee return. Since both four-element formulas were interpretable, the search for enough took seven years.

Again, the problem of finding a valid interlocutor was upsetting. The only thing the mujahideen could agree on was a refusal to negotiate, so, similar to South West African negotiations, Pakistan became the party dealing with the government, bringing in a new formula. The Soviet leaders who authorized the invasion was replaced in 1982 by Yuri Andropov, who opposed it, and he in turn was succeeded in 1985 by Gorbachev, who in 1987 warned the Afghan government that Soviet troops would be out in a year. (Bokhari 1995: 239). While negotiations turned to Pakistan versus Afghanistan, the mujahideen continued their position politics under the cover of the war. The result was instability nurtured by internecine conflict even after the Geneva agreement was signed until the Taliban arose in 1994 to fill the power vacuum created by intra-mujahideen conflict with a message of righteousness and clean government.

The next Afghan war began with the reemergence of the Taliban after its removal from governing as punishment for the 9/11 attacks on New York and Washington in 2001. The International Security Assistance Force (ISAF) responded to remove the insurgency, but as war-ending by force proved to be illusory, various attempts were made on negotiations by the government, the UN, and the US. President Ahmed Karzai offered negotiations throughout his second term until 2011; if there was a sense of a hurting stalemate, it was not mutual as both the Talban and the US rejected the idea of negotiations. Although individuals on both sides recognized that the conflict was at a draw, imposing heavy costs on both sides, to one side the Taliban represented a dangerous and unthinkable partner in negotiation and to the other divine promise overshadowed any need to negotiate. By the later 2010s, the Taliban was in control of over half the country, despite ISAF surges. Karzai's successor, Ashraf Ghani responded to a Taliban announcement that 'the issue cannot be solved militarily' with a proposal for negotiation in 2018. Again it is difficult to pinpoint when a feeling of. a mutually hurting stalemate was established; if it ever appeared it was fleeting and a one-sided feeling was seized by the other side as a time to press its advantage. When the surge under President Obama or an annual Taliban summer offensive did not produce either victory or offers to talk, the Tactical Question was continually brought up on (as far as we know) both sides but at different times.

There was never any. ripe moment or mutual and simultaneous feeling of hurt and stalemate. But while the military continued to claim the possibility of an eternal (or at least until the Afghan army was properly trained) presence of 2500 troops in the capital area, it was the US and allies, like the Soviets earlier, that were deeply hurt and stymied, politically (which the military did not understand) even if not militarily; the Taliban were never stalled not pained; there was always the possibility of a good offensive next spring for the good cause. The search for Enough took two years of off-and-on negotiation in Doha, during which the Taliban gradually acquired more and more territory. Secret talks were begun and failed in 2011; their initiation in Doha in mid-2018 was more promising, although continuously interrupted by sporadic violence. President Trump's announcement of US troop withdrawal changed the negotiating balance: on one hand, it declared a commitment to end the war, but on the other, it removed any bargaining chip the US had. In essence, the US said, 'we will withdraw if we can come to an agreement, and if we can't, we will withdraw'. It was clear which side felt a stalemate. In this situation the parties reached an agreement with the Taliban political leader in February 2020, for a pledge not to allow al-Qaeda in Afghanistan (the original purpose of the 2001 invasion) and an agreement to negotiate with the Afghan government, in exchange for full ISAF withdrawal by October 2021, leaving no means of holding the Taliban to its terms, which it did not. The way to end this war was to end it, by full unconditional withdrawal (except for those left behind), on 24 September 2021.

Ukraine

It is timorous to analyze negotiations in the War of Russian Aggression in Ukraine since at this writing (April 2022) it is not over. Yet three ends and the path to them are visible: either Western and Ukrainian fatigue and defeat, or Russian fatigue and defeat, or something in the middle. None of these is possible without negotiation, after the war has done its worst.

In the first three months of the war and in the seven years before that, negotiations were tried after the Russian infiltration of the Donbas (the two oblasts of Donetsk and Luhansk) in eastern Ukraine without providing acceptable results. Russian aims, personal to President Vladimir Putin, were unclear, ranging from absorption or all or half of Ukraine to full or symbolic annexation of Donbas like the other scabs of South Ossetia, Abkhazia, and Transnistria, within a broader goal of neutralizing Ukraine and withdrawing NATO armaments from neighboring countries. Putin declared that Ukraine never existed as a country and that Ukrainians should to eliminated. After a hybrid war in Donbas, France and Germany with the OSCE tried to mediate interim ceasefires, but despite some signatures in a succession of Minsk Agreements, the terms were not enough for Russia and too much for Ukraine.

In February 2022, following a series of speeches in Beijing laying out a new system of the world order of ultra-Realism dominated by a Russo-Chinese axis imposing the abjection of NATO countries and the liberal world system, Russia launched its invasion. When a blitzkrieg that should have taken Kyiv in three days bogged down with heavy losses imposed by fierce Ukrainian resistance and civil defense, Russia turned to the conquest of the whole Donbas region, which it had recognized as independent states, and the Black Sea coast of Ukraine; meanwhile it prepared a second assault on Kyiv to remove the shame of its previous defeat, while at the same time conducting genocidal civilian massacres to terrify the population.

Whether the war will end in a negotiated draw depends on the ability of the Russian army to perform better than in the first round and the ability of the Ukrainians to block and repel their efforts, posing then the Enough Question. In normal conflict conditions, a stalemate would bring up the conditions of ripeness and provide for a shared search for mutually enticing opportunity acceptable to both sides. In this case, however, the cost already exacted on Ukraine is so high as to outrank any question of compromise. Behind that, the brutality of the Russian tactics and vision has hardened Western solidarity in Europe. Yet the personal commitment of Putin is such that it eliminates the likelihood of compromise on his side, and Russian leaders and populace seem to support him. Chances of replacement leadership from the army or from the capital seem slim and lacking historic precedence. In the case of a Russian victory and the death of Ukrainian president Volodymyr Zelensky, negotiation would be over the terms of surrender and occupation with an imposed Ukrainian spokesman. In the other case, negotiations would encompass the Russian withdrawal, establishment of reparations, and eventually worldwide negotiation for the establishment of a new world system designed to prevent a future breakdown of the international system of norms and institutions. Peace depends on the war and how it ends. Negotiation is required in that ending and, even more creatively, in drawing up the architecture of a new world order.

Analysis

These five cases provide a good deal of clarity to the conceptual questions at the beginning. Wars are dichotomized (or pluralized) conflicts of force; when the parties begin to doubt that force will win for them and they share those doubts, they then can consider negotiations, and the confrontation broadens to include issues of Enough, not just relations of force. That decision may well be a bit waffled and unsteady, but its continuity depends on forward movement in the discussion and on reminders of the dangers of return to force (BATNA) if the parties were to fall backward. The more stable the conviction on both sides that neither can win militarily, the more likely a war-ending agreement becomes. That likelihood is not a certainty until the agreement is signed (and then implemented), but progress to the end does carry a certain predictive momentum toward a foreseeable outcome. In all three positive cases, an awareness of ripeness required a long conflict and mounting losses to be convincing. In the fake case, it appears that strong ideological convictions on both sides that 'we must win and dare not lose' prevented ripeness and led to an illusionary agreement that one side observed in haste and the exchange partner did not. In the still-open case, war-ending negotiations will depend on the outcome of the war to end; the outcome of an MHS is conceivable, either as a balanced stalemate of fatigue and unachieved goals or a one-sided stalemate with the loser still alive.

Since that decision and then forward progress requires a policy reversal on both sides, a valid spokesman is crucial. In all the cases, action on perceived ripeness and a reversal of former policy carried with it (or was carried by) a change in leadership on one or both sides. The change involved negotiations and dynamics that went on within each side, and it is here that the Toughness Dilemma and the Enough Question emerged in strength. Debates were settled by retirement through death or elections when persuasion failed. There has not been space to take these internal negotiations into account but they were crucial. Crucial they are too in the Ukrainian case, in scenarios that are easy to envision but difficult to predict. The great danger in this case is that it will be deemed necessary to give Putin a 'face-saving' sop from the already damaged and wounded victim of his aggression.

It is most important to note that in none of the accomplished cases did Enough require a change in the goals of the parties, only in the means to those goals. Goals could be redefined, but fundamentally remained unchanged. In many cases, the parties pulled back from the maximum expression of those goals to find a functional equivalent that met the criteria of Enough. FARC wanted agrarian reform and participation; Colombia wanted the end of violence and punishment. a change of leadership on both sides was necessary to bring out ripeness and to carry its conclusions into action. Both parties in South West Africa wanted the withdrawal of the opponent's troops; one party did win over the other on the issue of Namibian government but that victory was followed by a monumental change on the underlying issue of South African government. In Mindanao, the Moros wanted high autonomy; the Filipinos wanted an end to conflict without secession and so they invented a look-alike that satisfied both sides after they learned that submissions or independence were impossible. It is hard to deal with the complexity of the first Afghan war in these terms; in the second war, the Taliban's goals remained unchanged and the US fell back onto its goal of two decades ago on the removal of al-Qaeda and so they talked past each other and the Taliban got there first. The Taliban's BATNA was a takeover, which they got; the USA's was withdrawal (with which they threatened the Taliban during the negotiations!), which it got, in unseemly haste.

It is this characteristic that troubles Ukraine negotiations. In Ukraine, the parties have toyed with something between the two side's goals, in the Minsk negotiations, neither finding the result enough. Ukraine offered the functional equivalent of Russia's demand, in guaranteed neutrality with open external relations, not enough for Russia; Russia offered nothing to Ukraine except protectorate status and scabs. Now the cost has been raised so high and the egregious behavior of the aggressor dragged so low that a 'split-the-difference' or reframed goal type of outcome is difficult to swallow. Normal negotiations may indeed be a trap rather than an opportunity.

In the usual practice, if one side does not give up, and if continuing the war is deemed painful and pointless by both sides, the parties can look for another way to define and attain their goals. This requires some sense of a policy and leadership change, but it ends in trading in violence for some other goal achievements, thus breaking the original stalemate and cutting the unacceptable losses. That is the way of negotiation. Cases show that is possible, but in zero-sum conflicts defined as existential, the war has to be won first.

References

Bokhari, I.H., 1995. 'Internal Negotiations Among Many Actors: Afghanistan', *Elusive Peace: Negotiating an End to Civil Wars*, edited by Zartman I.W. The Brookings Institution, R.R. Donelley & Sons Co. USA.

Calahan, H.A., 1944. *What Makes a War End?*. Vanguard, New York.
Carroll, B., 1969. 'How Wars Ed: An Analysis of some Current Hypotheses', *Journal of Peace Research*, vol 4.
Coker, C. (ed), 1997. *War Endings: Reasons, Strategies, and Implications.* Special Issue, *Millennium.* XXVI 3.
Fixdal, M., 2012. *Just Peace: How War Should End.* Palgrave Macmillan, London.
Fox, W.T.R., 1970. 'How Wars are Ended', *Annals of the American Academy of Political and Social Science.* Nov 1, 21.
Hampson, F. and Zartmann I.W., 2015. *The Global Power of Talk.* Routledge, New York.
Iklé, F.C., 1971. *Every War must End.* Columbia University Press.
Kahn, H., Pfaff, W. and Stillman, E., 1968. 'War Termination, Issues and Concepts, Final Report', Hudson Institute, Croton-on-Hudson, New York.
Klingberg, F.L., 1966. 'Predicting the Termination of War: Battle Casualties and Population Losses', *Journal of Conflict Resolution*, June, vol 10, no 2, pp. 129–171.
Kreutz, J., 2010. 'How & When Armed Conflicts End', *Journal of Peace Research*, vol IIIL2, pp. 243–250.
Licklider, R., 1993. *Stopping the Killing: How Civil Wars End.* New York University Press, New York.
Phillipson, C., 1916. *Termination of War and Treaties of Peace.* Dutton, New York.
Rose, G., 2012. *How Wars End: We Always Fight the Last Battle.* Simon & Schuster, New York.
Taylor, A.J.P., 1985, *How Wars End.* Hamish Hamilton, London.
Zartman, I.W., 1989. *Ripe for Resolution.* Oxford University Press, Oxford.
Zartman, I.W. and Aurik, J., 1991 'Power Strategies in Deescalation', in Kriesberg L. and Thordon, S. (ed), *Timing and the De-escalation of International Conflict.* Syracuse University Press, New York.
Zartman, I.W., 2000. 'Ripeness: The Hurting Stalemate and Beyond', in Stern P. and Druckman, D. (eds), *International Conflict Resolution after the Cold War.* National Research Council.
Zartman, I.W., 2004. 'Mutually Enticing Opportunities and Durable Settlements', in Zartman, I.W. (ed.), *A Pioneer in Conflict Management and Area Studies.* Springer, New York.
Zartman, I.W., 2006. 'Ripeness Revisited: the Push and Pull of Conflict Management.', in Hauswedell, C. (ed), *Deeskalation von Gewaltkonflikten seit 1945.* Klartext.
Zartman, I.W. and DeSoto, A., 2010. *Timing Mediation Initiatives.* United States Institute of Peace.
Zartman, I.W. (ed), 2019. *How Negotiations End: Negotiating Behavior in the Endgame.* Cambridge University Press, Cambridge.
Zartman, I.W., 2021. 'Understanding Ripeness: Making and Using Hurting Stalemates', in MacGinty, R. (ed), *Contemporary Peacemaking.* Springer, New York.

13 Negotiating Peace

Damien Kingsbury

This chapter focuses on negotiated settlements, with particular reference to separatist armed conflict. The principles of negotiated resolution may also apply to civil war and can in theory apply to interstate armed conflict, especially in cases where conflict is limited or stalemated. The principles raised in this chapter are not exhaustive, being based on negotiated outcomes or attempted negotiated outcomes across a series of separatist and internecine armed conflicts in which the author had experience.

Although it is understood in principle, negotiations aimed at ending armed conflict can in practice occur during armed conflict or during a ceasefire. That is, the negotiation process cannot be understood to be separate from the war it intends to end. In a related sense, parties to negotiations will often try to impose conditions on the talks prior to their commencement. While it may be possible for a more powerful party to impose such prior conditions, more commonly this confuses the role of negotiations with that of conditionality. It is the role of negotiations to determine what conditions may exist.

The real loss of lives, physical destruction and impoverishment that is a manifestation of war are strong psychological influences on participation in negotiations intended to achieve peace. The stakes in a negotiation aimed at ending wars are extraordinarily high, the difficulties of finding an agreement to satisfactorily separate combatants and to permanently end fighting and addressing the drivers of the conflict each presents complex challenges. If 'War is nothing but the continuation of policy by other means', (Clausewitz 1993: 77) or, perhaps, a failure of policy as is often the case in relation to separatist wars then peace negotiations are the continuation of wars by bargaining. So too, the 'fog of war' (Clausewitz 1993: 35, 45, 47)—incomplete or inaccurate information set against high levels of fear, doubt and excitement (and often lack of sleep)—might also be applied to negotiation processes. The process therefore can be, for participants, 'troubling, profoundly challenging and deeply humbling' (Kingsbury 2006).

There are a number of principles of negotiation which tend to apply to greater and lesser extents across most, if not all, such processes. They include: preparedness to be at the table; political change, including of leaders

DOI: 10.4324/9781003317487-18

of the parties to negotiations (democratic or otherwise), and 'ripeness' of the moment; hurting stalemates; spoilers and monopolies of force; roles of third parties, soft and hard intervention or external pressure and inducements, and neutral mediation; negotiation as politics; negotiation as transaction; rethinking goals; the quality of negotiation teams; acknowledgement of respective grievances; building trust in the process; disarmament and reintegration; security sector reform; and guarantees.

There is, too, a broad equilibrium between relative balances of military capacity and political orientation, although there are exceptions and emphases which can change emphasis within the overall process. To illustrate, one party might have a military advantage but if it is also undergoing significant political change, e.g. democratizing or de-democratizing separately to the negotiation process, then this can also alter the nature and balance of negotiations. There is also the question of the extent of third-party intervention, support or withdrawal of support, militarily, diplomatically or economically. The question of technology also needs to be brought into such consideration, given it can also alter the relative balances, e.g. in asymmetric warfare, for types of outcomes. The way the various parts all contribute to a negotiated outcome means that the metaphorical chess board of negotiations is not of a conventional two dimensions, but played on three dimensions, significantly complicating what can already be a complex conventional format.

Some also argue that peace needs to be 'sequenced,' particularly if it is a 'democratic peace,' that it should build on a number of logical steps (e.g. Langer et al 2016, Joshi 2015). Such a sequence might include the creation of an absence of violence or ceasefire, security sector reform (Toft 2010: 19, 37), building consensus, transitional power arrangements, disarmament, demobilization and reintegration, employment creation, democratization training, forming political parties within an agreed overarching framework, holding elections and respecting outcomes (Bell 2016, also Galtung 1996).

Others suggest, however, that peace is not so formulaic and that while certain steps are useful or individually necessary, there is no pre-ordained pattern or consistently necessary process that needs to be followed (e.g. Tornquist 2011, Kersten 2011, Carothers 2007 on democracy). The experience of the author is that sequencing can be considered as a form of created 'ripeness' and that while there are some attributes that contribute to a greater likelihood of success their inclusion and ordering are helpful rather than strictly necessary.

Preparedness to Be at the Table

The first principle of negotiated conflict resolution is that the parties to negotiations must want to negotiate or feel sufficiently compelled to negotiate. In a related sense, the parties must also regard the achievement of a satisfactory negotiated peace as necessary or highly desirable, in order for them to persist during the more difficult parts of a negotiation process. In this, it is important to note that all of a negotiation process is difficult and some parts appear, from time to time, to be unsurmountable. Continuing to come back to the negotiating table and persisting with the process is the only way that such a process can achieve a positive conclusion.

Preparedness to be at the negotiating table is likely to be greater in the case of protracted conflicts, given that hopes for a short, sharp victory will have been abandoned. Similarly, a more intense conflict, in which one or both parties are waging a higher-level war with the goal of victory is less likely to see both parties being prepared to negotiate, given that one or both have committed their resources to the alternative path of peace through military victory. A less intense conflict, then, in which one or both parties have less commitment to total victory, is more amenable to a negotiated settlement.

There is, however, always an element of compulsion in negotiations; a need to be there, for one reason or another. The difference is when one or both parties decide the negotiation is either a better option or at least worth exploring as an alternative to war, without being unwillingly compelled to the table, for example by a third party.

Negotiation as Politics

Negotiations are first and last political processes. War is not just politics by other means, to paraphrase Clausewitz, but a failure of politics. Negotiations aimed at peace, then, are intended to restore politics to the centre of relations and to remove violent conflict. So, too, the goals of negotiating are intensely political, setting against each other competing sets of interests that are so oppositional as to have devolved into armed conflict.

Negotiations may therefore address strategic, economic issues or humanitarian issues. These are processes of political bargaining, based on political foundations, with political outcomes and conducted through the exercise of politics. In this respect, politics can be understood as the practice of power or capacity to assert one's will, against a competing party's power or capacity to assert their will, set against the cost of business as usual. In the case of negotiations towards ending the violent conflict, the circumstances that bring parties to the table are the inadequacy or failure of politics by other means.

The process of negotiations is the mechanics of how fundamentally oppositional interests can be reconciled, or finding an agreed framework for abandoning some competing interests on a basis of exchange or agreed compromise, while establishing a regulatory framework for the resolution of remaining acceptable differences or differences that do not undermine the integrity of an agreed outcome. That outcome, if it is to be achieved, will necessarily compromise claims to the exercise of complete power and will, in practice, result in some degree of distribution of power. In this respect, the outcome of negotiations should not aim for one side or another to win through the negotiation process what they could not achieve on the battlefield but, if 'win-win' is an optimistic cliché, then for each of the parties coming away from the process without excessively losing.

Political Change and 'Ripeness'

Reasons for coming to negotiations to end armed conflict may be varied but tend to fall into categories, including 'ripeness,' hurting stalemates and, less commonly, where parties are prepared to explore lateral options to address claims. A common reason for parties to an armed conflict agreeing to negotiate towards a settlement and for the more likely success of such negotiations is that the situation is 'ripe' for resolution (see Zartman 2008, Schrodt 2003, Zartman 2000). That is, the circumstances which have led to the conflict have changed and are more able to be resolved, such as reforms aimed at ending the drivers for conflict, changes of leadership or where parties to the conflict have arrived at a point where they can see alternatives to conflict and where a range of pressures against conflict and towards resolution have developed or increased. To illustrate, the Indonesian government's drawn-out commitment to sustaining a military presence in the occupied territory of what was Portuguese Timor (later Timor-Leste) following the impact of the Asian Financial Crisis on the Indonesian economy and the end of President Suharto's tenure in 1998 created an opening for resolution.

Through United Nations brokered negotiations with the territory's nominal legal power, Portugal, in May 1999 Suharto's successor, B J Habibie, agreed to allow the UN to oversee a referendum on whether the people of the territory wished to remain as part of Indonesia under a 'special autonomy' arrangement or whether they wished to be independent. Indonesia was undergoing fundamental political change at this time, including its own international recognized free and fair elections in June 1999, which helped facilitate the change of position on the future of the territory.

Such 'soft' international intervention, in which there are no armed UN-authorized personnel, may, however, not be sufficient to achieve its stated goals. Such was also the case in Moroccan-occupied Western Sahara (formerly Spanish Sahara) where, under the auspices of UN adjudication, in 1991 the Government of Morocco agreed to hold a ballot on the future of the territory for the following year but quickly reneged on that agreement. Morocco's withdrawal from the agreement was over the UN's disallowing inclusion in the ballot of Moroccans who had emigrated to Western Sahara since Morocco's 1975 invasion and incorporation, as well as pre-existing Saharwis. That left the UN's overseeing body, UN Mission for the Referendum in Western Sahara (MINURSO), in the sole role of monitoring a fragile ceasefire. Polisario leaders had warned that Morocco's unilateral withdrawal from the agreement for a ballot on self-determination would inevitably lead to a return to armed conflict.

Morocco has since refused to allow the inclusion of an independence option in any future ballot but instead opted for 'autonomy' (under Moroccan sovereignty) (Jensen 2011). The International Crisis Group (ICG) has since attempted to re-start negotiations between the Moroccan government and Polisario (ICG 2021). Polisario President Brahim Ghali said that Polisario forces would continue to attack Moroccan positions in the occupied territory unless MINURSO was given a clear mandate for a vote on a self-determination (Africanews 2021). The ceasefire broke down and armed conflict resumed in 2020 following an attempt by Morocco to extend its effective control of Western Sahara further south via a road link to Mauritania.

In cases where 'soft' intervention has failed and in particular where there is potentially or actually significant loss of life, the UN had retained the final option of military intervention under the paradigm of the 'Responsibility To Protect' (R2P) established in 2005. This paradigm has three components, the first of which is that states have a responsibility to protect their own citizens and, where that may be lacking, the international community has a responsibility to provide assistance and capacity building. However, where a state 'manifestly fails' to protect its own population (or attacks its own population) then the international community should take collective action to protect that population. This latter aspect includes providing mediation and establishing early warning mechanisms, economic sanctions and, under Chapter VII provisions of the UN Charter as determined by the UN Security Council, the use of force (Burke-White 2012).

Despite being an action of last resort, it is this last element has been that which 'R2P' has been most associated and which is most controversial (Deller 2012). As an aside, it was the UNSC-mandated military intervention in East Timor in 1999 which was one of the most successful, effectively textbook, examples of the use of Chapter VII use of force even though this event took place some six years before R2P was established as a formal model. 'Humanitarian intervention' differs from R2P in that it does not include R2P's preventative powers and relies solely on the use of force that may be undertaken unilaterally and does not rely on UNSC approval or legal cover. The distinction is essentially that humanitarian intervention is predicated upon a right to intervene (Weiss 2016: 13) whereas R2P is predicated upon protection of civilians.

Hurting Stalemates

Related to 'ripeness' may be a 'hurting stalemate' in which, after protracted conflict, victory appears to be unachievable but surrender sufficiently undesirable or lacking in compulsion. This is where the cost of continuing is substantially greater than the potential benefits. Zartman suggests that a ripeness as a result of a mutually hurting stalemate is based on cost-benefit analysis (Zartman 2008) in turn reflecting rational choice theory (see Scott 2000) in which self-interest is employed to produce the greatest benefits (or reduce the greatest harm).

This understanding is correct in the sense that parties to a conflict wish to maximize their gains while minimizing their losses. Where few or no gains are being made over an extended period but losses continue, there is a rational incentive to look for alternatives. However, 'gains' and 'losses' need to be understood in terms of not just their utilitarian value but also in terms of how they sit against subjective claims, including notions of state unity, national identity, self-determination and proto/nationalist aspirations and, not least, a desire for the cost of the conflict to date not to have been in vain. That is to say, one or both parties to a conflict may find they are unable to make headway and that costs continue to mount, but these may be borne against more nebulous and less quantifiable ideas, or understandings that on the face of it are not strictly rational.

One example of such an idea was that of World War One, when belligerents felt they had to honour already terrible loss by persevering with the war, thus ensuring further loss of life. Another example is the idea that 'to struggle is to win', where the value of conflict is in its continuation rather than its foreseen conclusion, or when success is measured by mere existence, no matter how diminished. Similarly, where conflict is driven by particular forms of zealotry (ideological, ethno-nationalist, religious), revenge (which in small wars is common, especially at the local level), where there is a high level of mistrust or, indeed, a dominant or 'infallible' political actor may prefer an 'irrational' course of action others see as foolhardy, a continuing will to bear 'irrational' costs may be difficult to overcome (see Eriksson 2011: 58–100 on assumptions in rational choice theory). Indeed, one might propose that if the costs and outcomes of wars were known in advance, there would be very much reduced enthusiasm for such conflicts to be entered into. It may be, then, that a dawning realization of the hopelessness of stalemated armed conflict is what brings parties to a process of negotiations, especially when both are already looking for a way out and negotiations may be face-saving device and thus worth entering into.

Parties are more inclined to accept negotiations or to progressing such negotiations when they have both been subject to a shock, which can act as a catalyst for an underlying desire to end conflict. The 'shock' in question may be anthropogenic, such as an economic crisis, or a natural disaster, for example that of the 2004 Boxing Day tsunami which devastated sections of South and Southeast Asian coastline and which contributed to, if not being the cause of, Government of Indonesia-Free Aceh Movement (GAM) negotiations to end three decades of conflict in the Indonesian province of Aceh. In this case, reflecting the 'soft intervention' approach noted above, the Government of Indonesia and GAM had actually agreed to negotiations to be mediated by the Helsinki-based Crisis Management Initiative two days before the natural disaster. However, the pressure on both parties created by the disaster, gave added impetus to the negotiations and provided compelling political, economic and humanitarian reasons for a successful resolution to GAM's claims for an independent state of Aceh and the Government of Indonesia's insistence that Aceh remains a province of Indonesia alongside other provinces.

On the part of the Government of Indonesia and in particular then newly elected President Susilo Bambang Yudhoyono, it had made a resolution of the Aceh conflict a key priority (Ziegenhain 2010: 132), not least as a means of reining in the Indonesian army (Tentara Nasional Indonesia, or TNI) which was then in the gradual (and still incomplete) process of coming under actual as opposed to nominal civilian control. This in turn was a key component of Indonesia's process of democratization, of which Yudhoyono, as a then senior army officer, was the principal architect (Rabasa 2002: 25–26, Kingsbury 2000). From Yudhoyono's perspective, ending the conflict would thereby make a substantial step towards ending the TNI's lack of accountability and its culture of impunity, as well as its meddling in domestic politics. By changing Indonesia's internal security needs, the TNI's self-defining role as 'guardian of the nation' would be diminished in a practical sense, while strengthening civilian oversight and helping to secure the country's still emerging plural political culture.

From GAM's perspective, after three decades of fighting and mounting casualties, particularly of civilians, and in the face of a relentless TNI campaign to crush the organization, it was beginning to realize that its hopes of a military victory were further away than ever and that, following the tsunami, a lack of international support had morphed into international opposition to an independent state of Aceh. It was therefore open to finding a new way forward.

The international community, in particular aid donors who had promised to assist Indonesia in its massive post-tsunami relief and reconstruction, made clear to both parties that aid would not be forthcoming without a resolution to the conflict but that sanctions against both would be applied. From an international perspective, too, the idea of finding a resolution to the conflict was driven by satisfying international partners by wanting to see Indonesia meet global norms and standards, which in turn fitted with its process of democratization. The international community offered a conventional 'carrot and stick' approach to supporting the process, promising to punish both parties if they did not achieve a resolution but rewarding them if they did.

Third Parties

It is the contention of this chapter that negotiations are more likely to be successful if they are mediated or facilitated by a neutral third party, by way of creating a controlled environment in which negotiations can take place, by averting 'blame' or points of contention not germane to the negotiating process or its outcome, and by creatively drawing on commonalities or areas of overlap as well as possible methods for resolving impasses.

In such circumstances, the participation, or soft intervention, of a neutral third party, may allow an opportunity for both parties to agree to a negotiation process without being seen to ask for such a process and thereby potentially betray weakness. Such was the case, for example, with the Government of the Republic of the Philippines-Moro Islamic Liberation Front (GRP-MILF) negotiation process, with Malaysia acting as the neutral mediator.

In this, a neutral mediator is critical if a negotiation process is to be accepted by both parties as an essentially fair process and worth entering into or persisting with. If the mediator is not neutral, or if there is otherwise a perception that the process of negotiations is not fair, then the aggrieved party will be disinclined to continue with it. In some of the literature on conflict resolution, a distinction is drawn between a mediated conflict resolution process and a negotiated conflict resolution process (eg Bercovitch 2007).

A mediated armed conflict resolution process will necessarily require concessions on the part of one or both parties to the conflict and achieving such concessions cannot be achieved other than through negotiation. To this end, while it may be possible (if more difficult) to negotiate a resolution of an armed conflict without mediation, a dichotomy between mediated armed conflict resolution and negotiated armed conflict resolution is a largely false and unnecessary one. A mediator needs to be able to sit above the positions being put by combatant parties and should also be well informed about the history and character of the conflict being mediated. To this end, a mediator will often need to rely on a team of similarly impartial advisers, county specialists and so on, as well as technical experts who can advise on the logistics of conflict resolution mechanisms such as economic incentives, the possible role of peacekeepers or peace monitors.

Negotiation Teams

The relative complexity and in particular the often-shifting terrain of peace negotiations, such as changing perspectives and strategies, the status of armed conflict, the influence of spoilers and so on, mean that parties to a negotiation process should also have a team of advisers. These are to assist with drafting negotiating points, providing intelligence and advice of scenario planning, drafting proposals, advising on strategy and helping ensure that the negotiating team is as best prepared as possible for the process they are undertaking. In this, being underprepared or potentially being taken advantage of may weaken one side in a potential negotiated outcome to the extent that they will find the outcome unacceptable and not be able to commit to it. Therefore, a well-advised and 'strong' negotiating team is not an impediment to negotiations but in fact necessary to help ensure the best possible outcome that is able to be agreed upon by the principal parties.

Critically, too, incorporating into overarching planning the often-shifting strategic ground that underpins the negotiation process. This is especially so if negotiations are undertaken while violent conflict continues during the negotiation process, which may be the case. It should be noted here that ceasefires are a temporary cessation of violent conflict and may assist but are not necessarily a given in negotiation process intended to achieve a peaceful resolution. To illustrate, the United States used its bombing campaign of North Vietnam as a bargaining chip during the early stages of the Paris Peace Agreement negotiations. In Aceh, the TNI continued its military campaign against the Free Aceh Movement/GAM throughout the Helsinki peace talks, despite a unilateral GAM ceasefire. Similarly, violence was part of the negotiating process during the Turkey-Kurd negotiations between 2012 and 2015 (Toktamis 2017). In this respect, the use of violence may be understood as a bargaining tool in the negotiating process, rather than as a violation of that process (Hoglund 2008: ch 2, 2011).

Regardless of incentives or potential punishments, in order to negotiate from a coherent position, parties to negotiations must each be unified, as well as having full authority to negotiate on behalf of their constituencies, with the agreed support of their constituencies; there is little point beating out the detail of an agreement if it then needs to go to a distant party in order to be accepted, modified or rejected. The issue of unity is especially important in relation to armed non-state actors (Ramsbotham et al 2016: 204) where representation is usually not tested through conventional processes such as democratic elections but is also critical where the state does not exercise full control over its military, as was the case in Indonesia.

In this respect, for negotiations to be able to achieve in practice any agreement in principle, parties to negotiations must also command a monopoly on their respective use of force. This parallels Weber's definition of the state as having a monopoly on the use of violence (1978: 54) but also including non-state actors having a similar monopoly. In an extended way, parties to negotiations need to be, and to be seen to be by their constituencies, as legitimate. That is, even if constituents do not agree with the detail or, indeed, the overall direction of negotiations, they accept that the negotiating parties have a right to negotiate and that the outcome of negotiations will be accepted. In this respect, legitimacy can be understood, to paraphrase Locke (1988: 104-6), as having consent to negotiate.

A monopoly on the use of force includes the ability to exclude or neutralize 'spoilers' or sub-parties which might have an interest in undermining either the negotiation process or any resolution achieved, as well as factional differences which might present competing claims of representation and to otherwise resist or overcome fragmentation. This is the opposite proposition of some 'peace' literature (Hampson 2007: 44), which instead argues for bringing spoilers 'inside the tent.'

To illustrate, military leaders or sub-leaders might have an interest in continuing conflict for the economic, political or status benefits it could bring, while competing ideological factions might have strongly opposing views to the nature or extent of possible concessions or benefits in a negotiated outcome.

For negotiations to proceed without the threat of being undermined or derailed, 'spoilers' must be able to be successfully neutralized or excluded from the negotiation process and also to be neutralized in any post-agreement environment.

Neutralizing spoilers depends on whether they are strong and hence kept out of the process, to be neutralized outside the negotiation process, or whether they are relatively weak and are able to be included in the negotiation process without derailing it. In either case, if spoilers are strong or not easily controlled by the respective parties, they may be successful in either undermining the negotiations process or the potential subsequent peace.

GAM also had established a monopoly of force in its own sphere, if in one case through the use of violence (the 2000 murder of Majles Pemerintahan/ Governing Council-GAM secretary-general, Don Zulfahri in Kuala Lumpur), while the government of Indonesia had to remove two senior TNI officers who were part of its negotiation team, to neutralize their attempts to wreck the negotiations process. Ahead of Round Three (of five rounds of talks), the two military 'spoilers,' Major-General Syarifudin Tippe and Admiral Widodo Adi Sutjipto, were removed from the Indonesian delegation (Kingsbury 2006: 74-5). While this did not automatically lead to a smooth negotiation process, it did remove from the talks the two most intransigent actors in the Indonesian delegation and thus made continuing and new problems less difficult to address. At this time, too, the TNI was weakened, relative to its previous political influence.

Further, the president who consolidated Indonesia's transition towards democracy, Susilo Bambang Yodhoyono, was a former senior TNI general and the senior member of the TNI's reformist faction, thereby with direct support from a major grouping within the TNI and able to assert his political authority over recalcitrant generals. In this, Yodhoyono demonstrated the proposition that political reform is best undertaken in coordination with 'softline' military officers, to have sufficient authority to carry reform through while also retaining the ability to control anti-reform elements (O'Donnell 1986: 40).

It is also worth noting that the Aceh peace agreement was largely achieved because Indonesia was consolidating its (still imperfect) democratic reform, while GAM was also open to a local democratic process. Much has been written about the idea of democratic dividends and democratic peace (Hobson 2017, Patapan 2012, Reiter 2012, Reiter 2010, Rassler 2005, Rosato 2005, Grayson 2003) and much has also been written about the failure of democracy to secure peace (Russett 1995). But the simple proposition that democracy is a method of managing political conflict within an agreed regulatory framework means that peace is more likely to be achieved and sustained when both parties to conflict resolution negotiations agree to a mutual democratic process (Jarstad 2016). Such an outcome guarantees that both parties retain open, plural political frameworks that can accept difference and disagreement while remaining within wider acceptable parameters. Such was the case with Aceh, even if Indonesia's own compromised democracy ultimately tested the fabric of the peace agreement a decade and a half later. So, too, one could identify weaknesses in Aceh's own democratic process, being more than merely procedural but perhaps not of an 'expanded' variety (Collier 1997).

Similarly, the MILF had established an effective monopoly of force over Muslim separatists on the island of Mindanao; other organizations existed but were excluded from the negotiations process and were marginalized in the field through persuasion or force.

So, too, the Liberation Tigers of Tamil Eelam (LTTE, or Tamil Tigers) had absorbed or eradicated competing claimants to representing the Tamil cause in Sri Lanka (Swamy 2006: 174–196, 199–244, 334-5) and had been able to participate in 2002 Norwegian-brokered talks as the sole Tamil representative and as an effective equal negotiating partner. However, between Tamil Tiger intransigence (and an unrealistic understanding of their military capacity) and a change of government in Sri Lanka to one that was not interested in negotiations, the ability of and interest by both parties in seeking a negotiated solution to their own long-standing conflict was shallow. Dispute arose over control of international aid allocated in response to the same 2004 Boxing Day tsunami that, ironically, assisted Aceh's peace process, eventually leading to a cessation of opportunities for further negotiations. On the LTTE side, this was exacerbated by the death in 2006 of its chief negotiator, Anton Balasingham, ahead of scheduled talks aimed at de-escalating tensions.

Around this time, an offer by the LTTE of a federal solution to the conflict, whereby the LTTE would agree to end fighting in exchange for the Tamil areas of Sri Lanka being self-governing within a federal system was withdrawn. The LTTE reverted to its original claim of demanding the creation on an independent Tamil state (Pirapaharan 2006). Armed fighting, almost inevitably, resumed, ultimately with fatal consequences for the Tamil Tigers and some 40,000 civilians with them (Darusman 2011: 41).

The West Papuan independence movement, which has been fighting a low-level guerrilla war against Indonesia since 1965, was a prime example of a movement so riven by factionalism that the Indonesian government was able to say, with a straight face, that it was prepared to negotiate but did not know who to negotiate with (it has never been prepared to negotiate West Papua's independence). In 2014, three of West Papua's independent movements, the West Papua National Coalition for Liberation (itself being an umbrella group for some 36 organizations), National Federal Republic of West Papua and the National Parliament of West Papua, coalesced to form the United Liberation Movement for West Papua. Even then, however, the National Committee for West Papua and the Free Papua Organization have continued to run separate and mutually undermining political and strategic agendas, so that West Papuans continue to press their claims through separate voices. As with

Timor-Leste, the Indonesian government and its military had and retain competing agendas in relation to West Papua, meaning that any political attempts at finding a solution to the West Papua problems tended to be undone by military and proxy militia responses undermining government initiatives (Kingsbury 2021: 196-8).

Similarly divided, in the remote hill tracts of north-east India, ethnic Nagas had been fighting an on-and-off campaign for independence since 1951, following their inclusion of their territory within India in 1947. While the Nagas have been fighting a guerrilla war with India, since 1988 Naga factions had also been at war with each other, not reaching an internal peace agreement until 2014. The negotiations that led to this agreement were conducted over successive days in a remote location in Nagaland, and followed a number of similar, if less focused, meetings in Chaing Mai over the previous three years. While the 2014 agreement allowed the dominant Naga faction, the National Socialist Council of Nagaland-Isak Muivah, to resume talks with the government of India in 2015, the Indian government was slow to progress these further negotiations, appearing to wait out the ageing NSCN-IM leadership (one of whom, Isak Chishi Swu, died in 2016) (see Kingsbury 2021: 174–179).

Ahead of Timor-Leste's ballot on self-determination, the Timorese Council for National Reconciliation (CNRT) acted as an umbrella group for all of the territory's pro-independence parties and groups. The Indonesian government, however, was divided, with President Habibie not being supported by all of his Cabinet and in particular by Security Minister Wiranto. This meant that the TNI felt emboldened to try to influence the outcome of the ballot through violence and intimidation, which succeeded in delaying the ballot by more than three weeks and, with the UN prepared to cancel the process until greater security had been achieved, came close to derailing it. CNRT, however, insisted that the ballot proceed, despite the violence, recognizing that if it was deferred the TNI would likely create an environment conducive to it remaining too unstable to be reconsidered.

Mediation

Assuming parties to a violent conflict agree to engage in negotiations, given there are fundamental and deep-seated points of contention—the respective parties have, after all, engaged in violent conflict over them—it is common for the parties to want to address the most critical issues first. This is a normal response, but it is an error. A good mediator will ask respective parties for a list of requirements necessary to be met in order to cease hostilities; those lists should be arranged in order of priority. The mediator should then reverse the lists so that the least important issues are addressed first.

For negotiations to be meaningful, both parties need to bring to the process a list of claims or items for negotiation. These should be structured in such a way as to clearly delineate, for each team's internal use, items that are able to be sacrificed at little cost, those that may be compromised on but in expectation of similar compromise from the other side, and those that are fundamental or only compromised on in ways which can preserve their core values, as well as to be able to take 'wins' back to constituencies. In this sense, peace negotiations are highly transactional.

Once talks start, it is critical that parties establish initial points of agreement and, by working from easiest to most difficult, it is more likely that agreement will be reached over points of least contention. The purpose of this approach is to build confidence in the process and trust between the parties and to help ensure that, if the process is to fail, it does so at the last hurdle rather than at the first hurdle. In particular, it is important for the mediator to emphasize achievements over failures (which can be set aside as 'temporarily unresolved') and to seek to reinforce positives over negatives.

By building confidence in the process and working on how to achieve agreement, even if on minor issues, can start to build momentum and, as was the case with the 2005 Aceh peace agreement, may achieve agreement on all other points but for the most contentious. In that situation, the last point—on whether Aceh could have its own political parties (Indonesian law stipulates that all parties must be national in composition)—was able to be agreed to in principle just after the talks process had formally ended and thus secured the overarching agreement.

Conflict resolution negotiations are also framed by the wider context of perceptions of 'national interest' and community perceptions of armed group claims, the role of intransigent or opportunistic political actors and media portrayals of respective claims. Although the negotiations occur within a confined space, the public relations 'war' can also have a significant bearing on processes and outcomes. How respective parties prosecute their arguments in the wider public domain can have bearing on how the international community will react or to whom it will lend support, as well as helping to bring more firmly behind the negotiation process civil society groups and other that might have felt marginalized by what are usually combatant-only processes.

In the compromise of negotiations, there also needs to be some acknowledgement of the legitimacy of grievances, and that respective claims should be accepted as nuanced rather than oppositional. In this, parties working together to resolve each other's claims strengthens trust as well as finding creative solutions to problems which challenge both sides. Parties to negotiations also need to understand what the claims or interests they bring to the process are able to be lost or modified and those which are core. In this, a mediator may have a role in guiding parties to identify core interests, to work through their priorities, to avoid zero-sum outcomes, to think creatively about compromises which leave intact core claims or interests while providing sufficiently satisfactory outcomes for non-core claims—a 'zone of possible agreement' (Ramsbotham et al 2016: 55).

Of critical importance, too, is for a mediator to ensure that language used by the parties is sensitive to respective foundational claims and that 'red flag' terms, such as 'independence,' 'unitary state' and so on are not used, lest they be perceived as a sign of intransigence or unbridgeable offence.

In this, it is important to step back from the conventional expression of claims and to unpack what such claims are intended to achieve; often the question is not about the core claim itself, such as 'independence' but what that is intended to achieve. If the goal is to produce the goods associated with 'independence', eg reduced repression, economic equity, religious freedom and so on, then by focusing on the goods rather than the overarching concept of 'independence' it may be possible to consider ways in which those goods can be achieved. Similarly, a claim of 'freedom' might be insurmountable but, if 'freedom' is unpacked to express what it is hoped it might deliver, such as removal of oppression, greater security or local decision making, those constituent attributes might be more readily able to be addressed. By doing so, the oppositional or confrontational heat may be removed from negotiations. In this way, 'zones of possible agreement' may be achieved by reframing core claims.

Assuming a new set of political arrangements can be arrived at, the mechanics of ending fighting usually include a return to barracks (the non-operational deployment of combatants) or their disarmament and demobilization. This is also commonly associated with the reintegration of former fighters, which is critical to ensuring that there is no longer a readily available armed force to overturn a negotiated agreement. It also helps ensure wider collective security from potential 'freelance' armed actors, most commonly as violent banditry. Reintegration implies more than just civilian clothes, but employment creation programs to help ensure former combatants are gainfully distracted from other potentially violent pursuits.

In many cases, violent conflict exists because, or is exacerbated by, poor military disciple and/or lack of civilian authority over the security sector (army, police, etc). A part of many peace negotiations will, then, focus on reform of the security sector. In states with strong security sectors but weak governments, this may lead to the derailing of the peace process or even the downfall of the government. However, where the government is relatively strong, particularly where it has a high level of public legitimacy, and where there are external guarantors of negotiated outcomes, security sector reform can be an important, indeed necessary, part of a negotiated peace (Toft 2010: 19–25, 26–38).

While it may be tempting to achieve a partial agreement, particularly where one or both parties see their goals being met but without a final commitment to peace, as CMI founder, former Finnish President and Nobel laureate, Martti Ahtisaari, was fond of noting, 'nothing is agreed until everything is agreed.' The purpose of this rule was to reduce the likelihood of incomplete agreements and to raise the stakes for achieving complete conflict resolution agreements.

Post-Agreement

A peace agreement is, of course, incomplete if its outcome does not succeed in maintaining peace. To that end, a final agreement, if achieved, should include mechanisms for sustaining peace. These should include some type of guarantee that commitments made are fulfilled, including a return to barrack of belligerent forces, a neutral law enforcement model and an impartial judiciary, the possibility of an external guarantor/s which has an interest in and commitment to a successful outcome (Toft 2010: 30–32) when the situation remains volatile, the option of neutral peacekeepers, a commitment to economic rebuilding, and continuing on ground mediation and trust and confidence building.

To this end, a mediator or mediating institution should have the ability to bring on external parties to act as guarantors to both the negotiation process. This is to provide a secure and neutral environment and, in relation to a possible resolution, by way of being able to fulfill short- to medium-term external commitments around aid and reconstruction, economic development, peacebuilding and human rights, as well as longer-term institutional arrangements such as ballots and bringing into being new political entities such as local representative bodies (Walter 1997). The negotiation parties should, of course, agree to these guarantors having a key role in maintaining short-term stability and longer-term peace-building assistance. A mediator may also summarize achievements for both parties (sometimes creatively or with license), even where the parties themselves do not recognize such achievements as being such.

If an achieved peace is to be sustained then there also needs to be post-conflict stabilization and development. Recognizing this as a critical issue in a post-conflict environment, and helping to ensure an equitable post-conflict outcome, post-conflict rebuilding and development are often critical elements of a negotiation process. In this respect, it is critical to address the drivers of armed conflict, in particular the grievances which may have led to such conflict.

Employment programs and/or cash incentives for ex-combatants are a common feature of efforts at reintegration. Experience has shown that young men whose adult experience has been that of war are rarely often sufficiently skilled in other areas and, upon demobilization, may resort to banditry or even seek new leadership in order to return to conflict.

Similarly, the poverty that drives principally young men to take up arms is a critical issue to be addressed, particularly if a negotiated resolution to armed conflict is to address the drivers of armed conflict and not just its symptoms (Jenkins 2013: 112–113, Addison 2016). As part of redress schemes designed to boost what are usually stagnant or declining economies, and for their humanitarian contribution, compensation for victims is often a sweetener in negotiated settlements, by way of acting as an inducement to one or both parties to assuage the hurt of their respective constituencies.

Given any agreement to end conflict cannot be implemented in its entirety immediately, it is important to have a period of transition, in which provisions that have been agreed to can be put in place, usually according to a predetermined timetable. The staging of implementation of the agreement will often be in a particular order, for example for armed conflict to cease, for combatants to return to barracks or cantonments, for the introduction of neutral peace monitors/keepers, disarmament, demobilization and reintegration, for economic and mutually acceptable post-conflict political process to be put in place, and for post-conflict monitoring to identify triggers for a potential return to conflict. This phasing is particularly important where trust is low (such phasing is distinct from conventional sequencing, which refers to implementing changed structural conditions prior to negotiations or prior to the implementation of a negotiated settlement).

In a post-conflict environment, there will often be strong demands for a judicial process to determine whether crimes have been committed in the course of the conflict. To this end, a successful violent conflict resolution process will often include an independent post-conflict judicial process and/or a truth and reconciliation process. Justice, especially retributive justice, and reconciliation are distinct entities and one may be preferred over the other; justice may be more immediately satisfying but reconciliation if in tandem with restorative justice, may have a better prospect of longer-term success. It is critical, therefore, that while justice must be done and be seen to be done, judicial processes should not be constructed in such a way as to alienate combatants from participation or, indeed, from honoring the terms of the peace agreement (Langer et al 2016).

Finally, while peace agreements are intended to be final, and despite the 'peace dividends' that might be expected from an end to conflict, there is nothing absolute about them and commitment to both parties respecting agreements is only as strong as the goodwill of their intentions or the capacity of external actors to impose a continuing peace. That is to say, the permanence of agreements is not guaranteed. One or both parties may act to revoke their commitment if they believe the agreement is not being fulfilled from the other side or if they believe their relative status has changed and they can again achieve their original objective through violent means. Like democracy, then, the continuation of a negotiated peace is only as strong as the commitment to it and that, if taken for granted or abused, it can easily fail.

References

Addison, T., Gisselquis, R., Nino-Zarazua, M. and Singhal, S., 2016. 'Needs versus Expediency: Poverty Reduction and Social Development in Post-Conflict Countries', in Langer, A. and Brown, G. (eds), *Building Sustainable Peace: Timing and Sequencing of Post-Conflict Reconstruction and Peacebuilding*. Oxford University Press, Oxford.

Africanews, 2021. 'Western Sahara: Polisario Front leader vows attacks on Morocco will continue', Africanews.con www.africanews.com/2021/10/17/western-sahara-polisario-front-leader-vows-attacks-on-morocco-will-continue// [accessed 27 October 2021].

Bell, C. and Zulueta-Fulscher, K., 2016. 'Sequencing Peace Agreements and Constitutions in the Political Settlement Process', International Institute for Democracy and Electoral Assistance, Policy Paper 13, Stromsborg, November.

Bercovitch, J. and Jackson, R., 2007. 'Negotiation or Mediation? An Exploration of Factors Affecting the Choice of Conflict Management in International Conflict', *Negotiation Journal*, vol 17, no 1, pp 59–77.

Burke-White, W., 2012. 'Adoption of the Responsibility to Protect', in Genser, J. and Cotler, I. (eds.), *The Responsibility To Protect: The Promise of Stopping Mass Atrocities in Our Time*. Oxford University Press, New York.

Carothers, T., 2007. 'The Sequencing Fallacy', *Journal of Democracy*, vol 18, no 1, pp 12–27.

Clausewitz, O von, 1993. *On War*, in Howard, M. and Paret, P. (eds), Everyman's Library, London.

Collier, D. and Levitsky, S., 1997. 'Democracy with Adjectives: Conceptual Innovation in Comparative Research', *World Politics*, vol 49, no 3, pp 430–451.

Darusman, M., Sooka, Y. and Ratner, S., 2011. *Report of the Secretary-General's Panel of Experts on Accountability in Sri Lanka*. United Nations, New York, 31 March.

Deller, N., 2012. 'Challenges and Controversies', in Genser, J. and Cotler, I. (eds.), *The Responsibility To Protect: The Promise of Stopping Mass Atrocities in Our Time*. Oxford University Press, New York.

Eriksson, L., 2011. *Rational Choice Theory: Potential and Limits*. Palgrave MacMillan, New York.

Galtung, J., 1996. *Peace By Peaceful Means*. Sage, London.

Grayson, K., 2003. *Democratic peace theory as practice*. Working Paper No 22, York Center for International and Security Studies, York University, Toronto.

Hampson, F., Crocker, C., and Aall, P., 2007. 'Negotiation and International Conflict', in Webel, C. and Galtung, J. (eds), *Handbook of Peace and Conflict Studies*. Routledge, Abingdon.

Hobson, C., 2017. 'Democratic Peace: Progress and Crisis', *Perspectives on Politics*, vol 15, no 3, pp 697–710.

Hoglund, K., 2008. *Peace negotiations in the shadow of violence*. Martinus Nijhoff Publishers, Leiden.

Hoglund, K., 2011. '"Negotiations Amidst Violence" Explaining Violence-Induced Crisis in Peace Negotiation Processes', *Interim Report IR-004002*. International Institute for Applied Systems Analysis, Laxenburg.

ICG 2021. 'Time for International Re-engagement in Western Sahara', Briefing No 82, International Crisis Group, Brussels, 11 March.

Jarstad, A., 2016. 'Democratization After Civil War: Timing and Sequencing of Peacebuilding Reforms', in Langer, A., and Brown, G. (eds), *Building Sustainable Peace: Timing and Sequencing of Post-Conflict Reconstruction and Peacebuilding*. Oxford University Press, Oxford.

Jenkins, R., 2013. *Peacebuilding: From concept to commission*. Routledge, London.

Jensen, E., 2011. *Western Sahara: Anatomy of a Stalemate*. 2nd ed. Lynne Rienner, Boulder.

Joshi, M., Melander, E., and Quinn, J., 2015. 'Sequencing the Peace: How the Order of Peace Agreement Implementation Can Reduce the Destabilizing Effects of Post-accord Elections', *Journal of Conflict Resolution*, vol 61, no 1, pp 4–28.

Kersten, M., 2011. 'The Fallacy of Sequencing Peace and Justice', *OpinioJuris* http://opiniojuris.org/2011/09/29/the-fallacy-of-sequencing-peace-and-justice/ [Accessed 14 November 2021].

Kingsbury, D., 2000. 'The reform of the Indonesia armed forces', *Contemporary Southeast Asia*, vol 22, no 2, pp 302–321.

Kingsbury, D., 2006. *Peace in Aceh: A Personal Account of the Aceh Peace Process.* Equinox, Jakarta.

Kingsbury, D., 2021. *Separatism and the State.* Routledge, London.

Langer, A., and Brown, G. (eds.) 2016. *Building Sustainable Peace: Timing and Sequencing of Post-Conflict Reconstruction and Peacebuilding.* Oxford University Press, Oxford.

Locke, J., 1988 (1688). *Two Treatises of Government.* Cambridge University Press, Cambridge.

O'Donnell, G. and Schmitter, P., 1986. *Transitions from Authoritarian Rule: Tentative Conclusions About Uncertain Transitions.* Johns Hopkins University Press, Baltimore.

Patapan, H., 2012. 'Democratic international relations: Montesquieu and the theoretical foundations of democratic peace theory,' *Australian Journal of International Affairs*, vol 66, no 3, pp 313–329.

Pirapaharan, V., 2006. 'Sinhala Leaders' Duplicity of War and Peace Has Left Tamils with No Choice but Political Independence', *Annual Heroes Day Statement*, Liberation Tigers of Tamil Eelam, Kilinochchi, Sri Lanka, 27 November.

Rabasa, A. and Haseman, J., 2002. *The Military and Democracy in Indonesia.* Rand Corporation, Santa Monica.

Ramsbotham, O., Woodhouse, T., and Miall, H. 2016. *Contemporary Conflict Resolution.* Polity, Cambridge.

Rassler, K. and Thompson, W., 2005. *Puzzles of the Democratic Peace.* Palgrave Macmillan, New York.

Reiter, D., 2010. 'Democracy and Peace', in Wade, J. (ed.), *How to Achieve Heaven on Earth.* Pelican, Gretna.

Reiter, D., 2012. 'Democratic Peace Theory', in Valelly, R. (ed.), *Oxford Bibliographies in Political Science.* Oxford University Press, New York.

Rosato, S., 2005. 'Explaining the democratic peace', *American Political Science Review*, vol 99, no 3, pp 467–472.

Rowen, J., 2016. 'Reconciliation in Divided Societies: Considerations for Timing and Sequencing', in Langer, A. and Brown, G. (eds), *Building Sustainable Peace: Timing and Sequencing of Post-Conflict Reconstruction and Peacebuilding.* Oxford University Press, Oxford.

Russett, B., Layne, C., Spiro, D. and Doyle, M., 1995. 'The democratic peace', *International Security*, vol 19, no 4, pp 164–184.

Scott, J., 2000. 'Rational Choice Theory', in Browning, G., Haleli, A. and Webster, F. (eds) *Understanding Contemporary Society.* Sage Publications, Newbury Park.

Schrodt, P., Yilmaz, O. and Gerner, D., 2003 'Evaluating "Ripeness" and "Hurting Stalemate" in Mediated International Conflicts: An Event Data Study of the

Middle East, Balkans, and West Africa', paper to the Annual Meeting of the International Studies Association, Portland, Oregon, USA, February.

Swamy, N., 2006. *Tigers of Lanka: From Boys to Guerrillas.* Vijitha Yapa Publications, Colombo.

Toft, M., 2010. *Securing the Peace: The Durable Settlement of Civil Wars.* Princeton University Press, Princeton.

Toktamis, K., 2017. 'The Peace that Wasn't: The Spectre of Violence during Negotiations between the Turkish State and Kurdish Mobilization, 2013–2015', paper to European Consortium for Political Research General Conference, Oslo, 6–9 September.

Tornquist, O., 2011. 'Dynamics of peace and democratization: Aceh lesson', *Democratization*, vol 18, no 3, pp 823–846.

Walter, B., 1997. 'The Critical Barrier to Civil War Settlement', *International Organization*, vol 51, no 3, pp 335–364.

Weber, M., 1978. *Economy and Society*. Roth, G. and Wittich, C. (eds). University of California Press, Berkeley.

Weiss, T., 2016. *Humanitarian Intervention.* 3rd ed. Polity, Cambridge.

Zartman, I.W., 2000 'Ripeness: The Hurting Stalemate and Beyond', in Stern, P. and Druckman, D. (eds), *International Conflict Resolution after the Cold War*. National Academy Press, Washington.

Zartman, I.W., 2008. '"Ripeness": the importance of timing in negotiation and conflict resolution', *E-International Relations*, 20 December. https://www.e-ir.info/2008/12/20/ripeness-the-importance-of-timing-in-negotiation-and-conflict-resolution/ [accessed 21 October 2021].

Ziegenhain, P., 2010. 'The Aceh Conflict during the New Order and the Following Democratization Process', in Gref, A., Schroter, S. and Wieringa, E. (eds), *Aceh: History, Politics and Culture*. Institute for South-East Asian Studies, Singapore.

14 Endless Wars, Perpetual Peacekeeping?

Adam Day and Charles T Hunt

Introduction

Since the end of the Cold War, UN-led peace operations have been a common component of the international community's response to large-scale civil wars (Guéhenno 2018: 185–96; Gowan 2018: 171–184; Howard 2018: 127–71). There is a significant and growing evidence base demonstrating a meaningful correlation between the presence of UN peace operations and conflict resolution, including empirical scholarship on the links between UN deployments and lower levels of violence, decreased risks of recidivism, and even longer durations of peace (Walter 2020: 1705–22). However, the bulk of these findings is based on relatively old data, employing longitudinal approaches where pre-2013 datasets form a significant part of the evidence base, and where more recent trends in both conflict and peace operations may not be captured. Moreover, newer data indicates that today's conflicts may be more intractable than twenty years ago, more likely to relapse, and more impervious to the kind of peace settlements a typical UN operation is deployed to support (Einsiedel 2017). Indeed, today's peace operations—frequently deployed in settings of active conflict and with few prospects for a durable political settlement—appear to be struggling to achieve the kind of meaningful and sustainable peace that earlier operations may have helped to bring about. Better understanding the contribution of contemporary UN peace operations, and how they are changing, can contribute to our broader understanding of how wars end.

This chapter explores the efforts of modern UN peace operations to end conflicts in settings with 'little or no peace to keep,' where active and deeply entrenched conflict drivers complicate any efforts to achieve the kind of successes highlighted in earlier waves of peacekeeping. Specifically, it draws on experiences in South Sudan, Mali, Central African Republic (CAR), and the Democratic Republic of the Congo (DRC), where UN peace operations are contending with new and emerging conflict trends that have thus far frustrated international efforts to broker peace. The chapter identifies several common factors that tend to work against peace

DOI: 10.4324/9781003317487-19

operations in these contexts, including (a) the lack of viable peace agreements with the key conflict parties; (b) deeply entrenched corruption and criminal networks that sustain conflict actors and disrupt typical UN state-building/stabilization efforts; (c) direct and indirect involvement by regional actors that undermine cohesive peace processes; and (d) a growing role of violent extremist groups in many settings, which often results in proscription rather than inclusion, allowing security-driven approaches to dominate. Our findings suggest that peace operations today are unlikely to achieve the kind of transformational change required to resolve many of today's conflicts, but can contribute to more sustainable stability if they find ways to innovate around different sizes and shapes of missions, build towards more realistic mandates and coherent strategies, partner with large financial institutions, and engage meaningfully with regional actors. Ultimately, as the UN builds towards a New Agenda for Peace, it should include a far more strategic look at the role of peace operations in ending wars.

The changing character of armed conflict

The past 20 years has witnessed a significant shift in conflict dynamics from the decade following the Cold War. Whereas the 1990s was a period of steadily declining rates of both intra- and inter-state conflicts, these trends have since reversed. Beginning soon after the terrorist attacks of September 11 2001 and the US-led invasions of Afghanistan and then Iraq, rates of intra-state conflict have risen steadily, reaching a level in 2015 that had not been seen since the 1980s. The rise of intra-state wars has led to a major increase in civilian deaths, including a sixfold increase between 2011 and 2017 (driven largely by the conflicts arising from the Arab Spring). Moreover, the return of intra-state conflict has been accompanied by higher rates of relapse following periods of peace: roughly 60% of conflicts that had been resolved in the early 2000s have since fallen back into a state of armed conflict (Gates 2016). Ironically, this high rate of relapse may be due to the success of international mediation: whereas the overwhelming majority of conflicts in the 1980s resulted in military victories, the past 20 years has seen roughly five times as many conflicts resolved via a mediated peace settlement. As Edward Luttwak (1999: 36–44) famously argued, military victory may provide a more sustainable (if brutal) outcome to civil war than a negotiated settlement.

These trends point to a conclusion that the conflicts of the past 20 years may be more intractable and less conducive to sustainable political settlement than ever before. One indicator of this is the average length of peacekeeping operation, which has steadily increased over the past two decades, reflecting the much longer time horizon for bringing peace to conflict-affected countries (Einsiedel 2017). Scholarship has identified three major factors behind the growing intractability of today's conflicts. First, the growing role of transnational organized crime has played a direct role in sustaining conflicts, undermining traditional efforts to pressure conflict parties, and has lowered the barriers to groups wishing to enter armed hostilities. The involvement of armed groups in the political economy of conflict—e.g. in marketing timber out of South Sudan, coltan from the DRC, or weapons across the Sahel—creates important channels for resources and resilient networks to perpetuate conflict (Tinti 2022).

Second, the role of violent extremist groups has dramatically increased in the past fifteen years, presenting new challenges to traditional conflict resolution approaches. Roughly 65% of major violent conflicts over the past ten years have taken place in Muslim majority countries (Walter 2017: 469–486), paralleled with a dramatic rise in the number and proportion of Salafi-jihadist fighters in today's largest conflicts (Jones 2014). These groups not only pose unique threats to civilians, but also complicate traditional mediation approaches due to their maximalist demands, isolation within the international legal regime (often via sanctions), and the strong tendency for Western actors to view them exclusively through a counter-terrorism lens.

Third, the growing willingness of international and regional powers to intervene directly in intra-state conflicts has tended to make wars more intractable and less conducive to mediated settlements. Research indicates that civil wars involving outside actors tend to last longer and be more deadly to civilians (Aydin 2011; Aliyev 2020: 630–650), as evidenced by longstanding instability in eastern Congo, the lack of progress on the political process in Syria and Yemen, and the most recent developments in Libya (Twagiramungu 2019: 377–91).

Taken together, these trends have meant that today's wars are far more destructive in terms of civilian lives and state institutions, less likely to be resolved via traditional inter-state forms of mediation, and highly susceptible to relapse. These trends apply directly to the settings where the UN has deployed peacekeepers, including the 20-year-old mission in the DRC, and what will almost certainly be longstanding missions in Mali, CAR, and South Sudan. While confronted with clear evidence that the landscape of conflict has changed, the UN has tended to resort to the same toolbox as it used in the post-Cold War period, employing what Gowan and Stedman (2018: 171–184) call the 'standard treatment' of mediation plus peacekeeping as the solution to most major conflicts today. The result has been the deployment of peacekeeping operations into settings involving continuing open hostilities (e.g. MINUSMA in Mali in 2013, MINUSCA in CAR in 2014), the continued operation of peacekeeping in settings without a viable peace process (e.g. MONUSCO in eastern Congo, UNAMID in Darfur), and/or the attempts to protect civilians caught up in the midst of an ongoing civil war (e.g. UNMISS in South Sudan). As the next section describes, this has created a set of recurrent challenges for peacekeeping missions which have limited the kinds of success outlined in some of the scholarship on the impact of peacekeeping.

Recurrent challenges in today's peacekeeping

Today's major UN peace operations are in many ways quite different to each other. Deployed to vastly different contexts where they relate differently to key conflict actors (particularly host governments), the missions in Mali, DRC, CAR, and South Sudan have certainly evolved their own characters. However, they share a number of common challenges that inhibit their efficacy in making and/or implementing peace, including: (a) the absence of viable peace agreements involving the major conflict actors; (b) competing priorities, particularly when it comes to the protection of civilians imperatives; (c) dependency on a fragile or highly eroded consent of governments and other parties to conflict, often limiting the utility of the use of force to achieve key mandate priorities; (d) the ability of transnational actors to actively undermine peace prospects and sustain conflict; and (e) the disconnect between the expectations of a mission's lifetime and the reality of the time it takes for the kinds of institutional transformation needed for sustained peace.

'No (viable) peace to keep'

The ability of UN peace operations to chaperone conflict-affected societies from war to peace is severely hampered by their deployment to settings where viable political processes that include the key parties to conflict do not exist. Sometimes they are wholly absent or moribund. For example, large stabilization missions were deployed to Mali and CAR before any comprehensive peace agreements were in place (though agreements covering some of the conflict actors were subsequently struck). In South Sudan, after supporting the independence of a new country, the outbreak of brutal civil war in 2013 led to a 5-year period where the mission was caught in a holding pattern while efforts to strike a peace agreement were primarily led by others (Hunt 2020). In the DRC, a 2002 peace deal that helped to end the country's long civil war expired only a few years later, after which a 20-year period has seen a proliferation of armed groups and conflict actors that are not constrained by any national-level peace process (Stearns 2017). In these cases, there has literally been 'no peace to keep'.

Where peace agreements do exist, they have often been unlikely to constrain the full range of conflict actors in a given setting. While a peace agreement was struck in Mali in 2015, it explicitly excluded jihadist groups and focused on the groups that had led the rebellion in the northern region of the country, ultimately failing to prevent the shift/spread of violent conflict to the central regions in the following years (Gorur 2020). Between 2015 and 2018, a key objective of MINUSMA became the need to shift resources and attention to central Mali, addressing a range of security threats from groups outside the 2015 agreement. Similarly, in South Sudan, the 2015 agreement between President Kiir and former Vice-President Machar failed spectacularly, in large part because it did not include a range of other groups in the southern Equatoria states, and elsewhere (Day 2019). While the subsequent 2018 peace agreement does nominally include these groups and thus offers UNMISS a more comprehensive basis upon which to operate, there are signs today that the fragile peace process may again disintegrate (Ero 2021).

The lack of a peace agreement does not doom a peacekeeping mission per se. Indeed, there are instances where a peace operation has helped to ripen a conflict to the point of a more inclusive political agreement, such as in CAR in 2019 and the 2018 Revitalised Agreement on the Resolution of the Conflict in the Republic of South Sudan. However, partial agreements, or settings in which the bulk of conflict actors are not constrained by an underlying political process (such as the DRC today) exhibit a resistance against the kind of progress that is often envisaged in Security Council mandates, often miring peacekeeping missions in quagmires without a clear exit (Gowan 2015: 39–46). Moreover, as our research has demonstrated, the absence of a peace agreement has often meant missions have been sidelined politically, forced instead to focus on the operational tasks of protection of civilians without the opportunity to shift a conflict towards more sustainable peace (Day 2020).

Competing priorities and POC

The emergence of the Protection of Civilians (PoC) as a centre of gravity across the UN's peace operations portfolio has been essential to saving lives and prioritizing people-centred approaches (Hunt 2019: 50–81). However, in some instances it has also led to distractions from other mission priorities and distortions of some intended objectives of missions (Day 2021a: 661–688). In several current missions, the centripetal force of PoC has drawn mission focus and scarce resources away from delicate political processes, at times complicating the positioning of the mission as an impartial actor. For example, during the 2016 constitutional crisis in the DRC, MONUSCO played an important supporting role for an AU-led attempt to bring all political parties together around an agreement for national elections. During the same period, the mission was conducting joint offensive military operations against the Allied Democratic Forces, an armed group operating in an increasingly pro-opposition area in eastern Congo. Population surveys at the time indicated a deep mistrust of MONUSCO in these areas, complicating the mission's attempts to project an identity of impartiality to the parties in Kinshasa (Stearns 2016). At one point in the mission's discussions of this challenge, a senior UN official captured the problem succinctly, 'How can we talk to the opposition parties here in Kinshasa while our troops are bombing their people in the East?' (Quoted in Day 2022).

Furthermore, the way in which missions have interpreted and implemented the PoC mandate has created situations where UN impartiality is seen to be compromised and missions are no longer viewed as an impartial arbiters. A clear example of this was in South Sudan where the creation and maintenance of PoC sites—hosting key communities associated with the opposition—generated friction with the host government (Day 2019). Similar dynamics have been noted in CAR, where the robust PoC approach has raised the possibility that the mission could be considered a party to the conflict (Maganza 2018).

As PoC has become the overriding priority for all major peacekeeping operations, our research has explored a worrying disconnect between protection and the overarching political objectives of the mission. While the growing practice of mission-wide PoC strategies indicates an attempt to approach protection in a more holistic manner, the strong gravitational pull of militarized forms of protection and the urgent need to protect those facing imminent risks, has fed a continuing tendency for PoC strategies to become divorced from deeper, longer-term objectives of sustainable peace (Day 2020).

The limitations of Consent

From the outset of peacekeeping over 70 years ago, operations have relied on the consent of the main parties—in particular the host government—for their mandate implementation. While this was a relatively straightforward issue in early peacekeeping missions involving interstate disputes, the evolution of peacekeeping to address instances of ongoing civil war has generated a growing set of challenges around consent. Specifically, where a government is involved in ongoing hostilities against a range of armed groups, some of which may be designated terrorists or otherwise targeted for offensive use of force, the UN can be pulled into a complex and often counterproductive set of relationships around consent.

On one hand, the need to maintain consent with a belligerent host government can present serious challenges to a mission's mandate implementation. In South Sudan, for example, the outbreak of civil war in 2013 meant UNMISS was physically positioned between government forces and hundreds of thousands of civilians, needing to shield them from attacks and facilitate access for humanitarian workers. While the mission maintained national-level consent to continue its presence in South Sudan, the day-to-day reality looked far less collaborative. Indeed, in 2016 government forces overran UNMISS bases and killed civilians within UNMISS PoC sites (CIVIC 2017). Speaking of consent in such circumstances reduces the term to nearly nothing, and certainly resonates with Johnstone's (2011: 168–182) concept of 'deteriorating consent.'

More broadly, consent often demands that the UN operate in close collaboration with government security forces, even where they are actively engaged in operations against rebel forces. In eastern Congo, for example, MONUSCO has engaged in joint offensive operations with the government against several armed groups for the past ten years. Similarly, MINUSCA tends to operate in tandem with the Central African Republic government and MINUSMA in direct support of the Malian security forces. While this may seem logical from the perspective of the UN's mandates to extend state authority and stabilize conflict-affected areas, in practice it may well render the UN de facto party to the conflict from the perspective of many communities in these settings (Bellamy 2015: 1277–1298).

Finally, these dynamics around consent have generated a dangerous trend in peacekeeping in which difficult governments may instrumentalize peacekeeping, obstruct operations, and resist more transformative change that might end conflict cycles. In Mali, for example, the government delayed and obstructed important provisions of the 2015 peace agreement, leaving the mission with few options to advance the process. In the DRC, the Kabila government expertly manipulated the UN peacekeeping operation for decades, bolstering its own authoritarian tendencies while superficially supporting the state-building work of the mission (Billerbeck 2019: 698–722). In CAR and Mali, the arrival of the foreign-backed Wagner group on the invitation of the host government has meant the UN has become embroiled in situations involving state-aligned militias that operate simultaneously as an arm of the state and an armed actor in the conflict.

In sum, the traditional application of consent as a necessary precondition of peacekeeping has become complicated, less likely to position the UN as an impartial actor in ongoing civil wars, and more likely to become a tool of manipulation by parties to the conflict.

Not in it for the Long Haul

As a result of the above challenges, today's UN peace operations do not appear to be well-positioned to oversee the sort of fundamental change required to transform conflict dynamics and lay the foundations for sustainable peace. Indeed, even as the prospects for relapse and recurrent civil wars have gone up, the ambition for peace operations has continued to grow. Today's UN peace operations are commonly mandated to support—sometimes lead on—the implementation of transformational institutional changes such as constitution-drafting, security sector reform, national rule of law reforms, and extension of state institutions into the peripheries of war-torn countries. At the same time, they are expected to prioritize protection of civilians in contexts where the very state actors they are meant to be supporting are often complicit in the violence. The result is what Gowan has referred to as the peacekeeping 'quagmire,' missions that become stuck in endless protection tasks as cycles of conflict are only partially addressed by a series of failed peace processes.

While it is clear that many of these settings will only emerge from recurrent cycles of conflict if they are able to address those deeper structural issues of inequality, underdevelopment, and lack of inclusive forms of governance, the ability (or even the appropriateness of peace operations to deliver such transformative change is very much unproven. This is in part due to the long time-horizon for implementing such national institutional reforms, which the World Bank has estimated at roughly 40 years for conflict-affected states (World Bank 2011). In the DRC, for example, efforts by MONSUCO to reform the criminal justice apparatus have been occurring for 20 years in partnership with various bilateral donors (Day 2021b). Yet, the Congolese court system and police are still only present in a limited number of locations and seen as legitimate and effective in even fewer. In CAR, the behemoth task of building—essentially from scratch—internal security forces that can deliver on a social contract across the country has proven so far beyond the capabilities of MINUSCA that there is little to show for seven years of trying.

This points to a longstanding debate about where peacekeeping ends and peacebuilding begins. Most recently, the AU/UN Hybrid Operation in Darfur (a mission that failed to deliver a comprehensive peace agreement after fifteen years) transitioned into a special political mission designed to help the Sudanese Government transition into 'peacebuilding mode.' This was seen by some as part of a broader likely trend of 'downsizing' peacekeeping into smaller, more political presences, more focused on root causes and politics, less on the more ambitious state-building and stabilization tasks of the bigger peacekeeping missions (Gowan 2020). Indeed, the well-known mantra of the High-Level Independent Panel on Peace Operations (HIPPO) report is that peace operations should move to a 'spectrum' that allows them greater flexibility for bespoke mandates responding more realistically to particular situations (UN 2015). However, despite many predictions of the end of large peacekeeping, it persists as a significant response to violent conflict; and, despite the call for more tailored mandates, they continue to emphasize robust use of force and stabilization approaches in complex settings like Mali, CAR, DRC, and South Sudan.

Reinventing the Flat Tyre?

The most recent package of reforms designed to fix these ills—the Secretary-General's Action for Peacekeeping (A4P) initiative—reflects these challenges but has thus far offered little to overcome them. Endorsed by over 150 states, the 2018 declaration tied signatories to seven key commitments (UN 2018a). One of these was advancing political solutions/enhancing the political impact of peace operations and supporting Security Council positions through their individual actions, including development, trade and military policies (UN undated). Efforts to implement A4P(+) have led to a range of technical improvements, better training for peacekeepers, a more coherent approach to monitoring and evaluation, and even some advances in the mobility of peace operations (UN 2019). These achievements should be acknowledged as important responses to the challenges of peacekeeping in complex environments.

However, the deeper challenges facing UN peacekeeping today remain largely unaddressed by incremental improvements to the same set of practices.

Better training of peacekeepers may help to change popular perceptions of underperformance in some settings but is very unlikely to address the fact that the use of force alongside state actors in today's civil wars tends to undermine the UN's impartiality. Improved monitoring and evaluation via the Comprehensive Performance Assessment System is an important step towards understanding the impact of operations on the ground but is unlikely to help peacekeeping overcome the root causes of structural inequality, venal state institutions, and underdevelopment that have allowed violence to remain endemic in many of these settings for decades. And thus far, the A4P initiative does not seem to have identified a solution to the fundamental problem of building regional and international consensus around peace processes. The withdrawal of French forces from Mali, continued fractures within IGAD over South Sudan, failure of SADC to cohere in support of eastern DRC, and deep divisions between the members of UN Security Council over the role of foreign fighters in CAR all in fact point in the other direction. This points to a concern that incremental efforts at reform may be likened to reinventing a flat tyre, pouring resources into a set of responses that are unlikely to advance the UN much down the road to sustainable peace.

A way forward

There is no immediate fix to the challenges facing peacekeeping today. The harsh reality is that Member States will continue to deploy the UN into settings of intractable conflict involving actors who are willing to pursue violence, repression and terrorism to achieve their ends. And on many fronts, UN peacekeeping has evolved important new tools to increase the effectiveness in these settings, including improved intelligence capacities, a more dynamic approach to POC, more nuanced human rights due diligence, and more holistic approaches to the use of force. We do not propose throwing these growing babies out with the bathwater. Instead, we here offer what we believe is missing in the Action for Peacekeeping initiative: a strategic-level set of transformations that will be necessary for the UN to evolve to meet today's conflict dynamics and ultimately become better at helping to end wars.

Get the HIPPO out of the Mud

One of the most important recommendations from the HIPPO was for the UN to think of a spectrum of UN interventions, rather than continue on with the binary of peacekeeping versus everything else. The 2018 reform of the peace and security pillar at HQ and the Action for 'Peacekeeping' have if anything reified this binary and made it more difficult to think creatively about the scope of options available for different conflict settings. Afghanistan has offered a moment of reflection for the UN to consider how to address a major transformation in a conflict environment. But more broadly the UN should ask deep questions about settings like Mali, where fracturing international and regional support may offer few points of traction for the current peacekeeping/stabilization approach. Settings like Haiti, which has seen widely different peace operations over the past 20 years, offer potential points along the HIPPO's spectrum, pointing to the possibility of 'all police' missions. And more creative still, the rise of cyber-driven forms of conflict could give rise to a new category of 'special technical mission' with a very different set of capacities (Dorn 2016; 2017: 138–146). The main point is to re-energize the important discussion started by the HIPPO and break down the paradigm that has kept peacekeeping producing the same stabilization missions over the past 20 years.

Independent Basis for Mandating

It has been two decades since Lakhdar Brahimi famously called on the Secretariat to tell the UN Security Council what it needs to know, not what it wants to hear (UN 2000). Yet today's peacekeeping practice appears to violate that principle as a matter of course. Peacekeeping mandating processes rarely acknowledge strategic failure, almost never open themselves to needed reorientations in the face of fundamental changes, and instead seem to continue to adorn the same Christmas tree mandates year after year. Twenty years of UN peacekeeping in the DRC has seen a continual growth of mandates around SSR and DDR, despite their near total lack of progress. The growth of independent strategic assessments has demonstrated the value of outside views (Forti 2021), but too often the UN Secretariat controls the narrative of these reports, pushing them ineluctably towards some version of what already exists on the ground.[1] Instead, the Security Council could include in its mandates the need for periodic, fully independent assessments of missions, used to inform the deliberations on future mandates.

Turn Missions on Their Heads

Peacekeeping missions are often deployed at moments of high risk to civilians and delicate junctures in peace processes, where the exigency to get boots on the ground may appear to outweigh many other concerns. This was the case in the 2006–2007 negotiations over UNAMID in Darfur, the 2015 mandate for MINUSMA in Mali, and the shift of mandate for UNMISS in South Sudan in 2013. The result is a set of mandates for peace operations that may not reflect the deeper challenges facing conflict-affected societies, the real prospects for achieving sustainable peace, and the ability of a peacekeeping mission to contribute to meaningful change. Instead, we propose that mandates be somewhat turned on their heads, with the Security Council issuing short, preparatory mandates for UN personnel to scope the realm of the possible, reporting back based on time in the field on what a realistic mandate might achieve. As we have argued elsewhere, this would not only achieve more realistic mandates but could go a long way towards building a stronger basis for consent with the host authorities (UNU-CPR 2017).

Financing Peace

Today's conflict dynamics are characterized by weak and/or predatory governance regimes, places like South Sudan, CAR, Mali and the DRC where patronage networks have come to dominate the political and economic spheres. The solutions in these settings are both political and economic, but peacekeeping is currently set up poorly on both of those fronts. One transformation could be to link peacekeeping mandates to long-term financial structures, multi-decade support from actors like the World Bank, International Monetary Fund and major donors. This would not only give the UN leverage via conditional finance and much larger sums of financial support but also demonstrate that the UN was engaged in the kind of multi-decade support needed for structural transformation. As laid out in the seminal UN/World Bank 2018 Pathways for Peace report (UN 2018a), a strategic partnership between the organizations could help address deeper questions of inequality and exclusion that are now widely seen as drivers of conflict.

Enhancing Strategic Coherence

Well-planned PoC approaches can bolster political processes, increasing confidence amongst conflict-affected populations in peace outcomes, enabling the building of trust amongst former belligerents, and reducing the risks of future violence. However, as discussed above, in practice protecting civilians and pursuing political strategies have at times been in tension. The gravitational pull of the PoC imperative can have unintended negative consequences on the rest of the mission and impede progress on conflict resolution and bringing fighting to an end. The UN has dramatically improved its mission-wide PoC strategizing and is grasping the nettle of the 'primacy of politics' with more meaningful political strategies. What is needed now is greater coherence between these overly siloed approaches. To address this incoherence the peace operations system needs to do more to align PoC strategies with political strategies and work towards more mutually reinforcing articulation of these two defining aspects of contemporary peace operations. Moreover, these two tracks must be incorporated into comprehensive and more coherent overarching mission strategies.

Towards Regional Prevention

As described above, many of the challenges facing UN peace operations today stem from transnational factors: illicit criminal networks, violent extremist groups, and troublesome regional powers, which all fuel conflict in ways that are poorly addressed via a traditional state-centric peacekeeping operation. The UN has taken major strides to building greater regional conflict prevention efforts, bolstering entities like the UN Office for West Africa and the Sahel, creating new regional structures like the Special Envoy for the Horn of Africa, and investing in a range of Regional Collaborative Platforms meant to coordinate UN actors across regions. This regional mindset could become far more part of the Security Council's approach to peace operations, and could become much more linked to the work of peace operations.

Conclusion

In September 2021, the Secretary-General issued a landmark report entitled Our Common Agenda, laying out an ambitious vision for evolving the multilateral system to address the many challenges facing the globe today. The report only mentions 'peacekeeping' three times, all in the context of funding modalities for the organization (UN 2022). Indeed, the 'New Agenda for Peace' initiative announced in the report makes no mention whatsoever of peacekeeping. This is a far cry from Boutros Boutros-Ghali's 1993 Agenda for Peace, which placed peacekeeping as one of the core tools of conflict management in the UN system (UN 1993). To a certain extent, Our Common Agenda appears to reflect the emerging priorities of the Organization, which have shifted heavily towards peacebuilding and so-called 'upstream prevention,' focused more on development-driven approaches to fragile, conflict-prone states. But there is a risk that the lack of strategic thinking around peace operations will stunt their evolution and lead to a set of practices even more out of synch with today's conflict trends, and missions in quagmires that are even stickier than now. Indeed, the war in Ukraine is likely to prompt a hardening of positions instead of the kind of innovative thinking needed for major change. Rather than relegating peace operations to the kind of incremental 'plus-plus' approach of the Action for Peacekeeping initiative, we here call for a deeper strategic process of thinking through how one of the UN's most important peace and security tools can be reinvented to be more effective at helping to end today's wars.

Note

1 It is worth noting here that a 2018 independent strategic review led by former SRSG Ellen Löj had initially pointed to the possibility of reducing the mission to a special political mission, but this recommendation was removed by the UN Secretariat before it was presented to the Security Council. See, https://minusma.unmissions.org/sites/default/files/180606_sg_report_on_mali_english_.pdf
https://walterdorn.net/pdf/Smart-Peacekeeping_IPI-Rpt_Dorn_13July2016.pdf

References

Aliyev, H., 2020. 'Pro-regime Militias and Civil War Duration', *Terrorism and Political Violence*, vol 32, p. 3.

Aydin, A. and Regan, P.M., 2011. 'Networks of third-party interveners and civil war duration,' *European Journal of International Relations*, vol 18, no 3, pp. 573–597.

Bellamy, A.J. and Hunt, C.T., 2015. 'Twenty-first century UN peace operations: protection, force and the changing security environment.' *International Affairs*, vol 91, no 6.

Billerbeck, S Von and Tansey O., 2019. 'Enabling autocracy? Peacebuilding and post-conflict authoritarianism in the Democratic Republic of Congo.' *European Journal of International Relations*, vol 25, no 3, https://journals.sagepub.com/doi/abs/10.1177/1354066118819724

CIVIC, 2017. Center for Civilians in Conflict, Washington, DC, https://civiliansinconflict.org/wp-content/uploads/2017/09/civic-juba-violence-report-october-2016.pdf

Day, A., Hunt, C.T., Yin, H., and Kumalo, L., 2019. 'Assessing the Effectiveness of the UN Mission in South Sudan,' Norwegian Institute of International Affairs.

Day, A., Gorur, A., Holt, V.K., and Hunt, C.T., 2020. *The Political Practice of Peacekeeping*. New York / Washington, DC, UNU-CPR / Stimson Center, https://collections.unu.edu/eserv/UNU:7791/OverarchingPaper.pdf

Day, A. and Hunt, C.T., 2021a. 'Distractions, Distortions and Dilemmas: The Externalities of Protecting Civilians in United Nations Peacekeeping.' *Civil Wars*; Kullenberg, J. N. 'Overlapping Agendas and Peacekeepers' Ability to Protect', *International Peacekeeping*, vol 28, no 4.

Day, A. and Caus, J., 2021b. *Rule of Law and Sustaining Peace: Towards More Impactful, Effective Conflict Prevention*. UNU-CPR, New York.

Day, A., 2022. *States of Disorder, Ecosystems of Governance: Complexity Theory Applied to UN Statebuilding in the DRC and South Sudan*. Oxford University Press, Oxford.

Dorn, A.W., 2016. 'Smart peacekeeping: Toward tech-enabled UN operations' *Providing for Peacekeeping No.13*. International Peace Institute, New York, https://walterdorn.net/pdf/Smart-Peacekeeping_IPI-Rpt_Dorn_13July2016.pdf

Dorn, A.W., 2017. 'Cyberpeacekeeping: A New Role for the United Nations' *Georgetown Journal of International Affairs*, vol 18, no 3.

Einsiedel, S Von et al, 2017. *Civil War Trends and the Changing Nature of Armed Conflict*. United Nations University Centre for Policy Research; ACLED data; Uppsala Conflict Data Program, Tokyo.

Ero, C. and Boswell, A., 2021. 'South Sudan's Dismal Tenth Birthday: The World's Youngest Country Needs an Overhaul', *Foreign Affairs*, July 9. https://www.foreignaffairs.com/articles/africa/2021-07-09/south-sudans-dismal-tenth-birthday

Forti, D., 2021. 'Learning From Recent Independent Reviews of UN Peace Operations' *Global Observatory*. International Peace Institute, New York, https://theglobalobservatory.org/2021/04/learning-independent-reviews-of-un-peace-operations/

Gates, S., Nygård, H.M. and Trappeniers, E., 2016, 'Conflict Recurrence,' *Conflict Trends*, vol. 2, February. PRIO, Oslo.

Gorur, A., 2020. *The Political Strategy of the UN Peacekeeping Mission in Mali*. United Nations University, New York, https://www.stimson.org/wp-content/uploads/2020/09/6-PeaceOps-2020-1245-Mali.pdf

Gowan, R., 2015. 'The Peacekeeping Quagmire.' *Georgetown Journal of International Affairs* (Summer/Fall), https://www.jstor.org/stable/43773693

Gowan, R. and Stedman, S., 2018. 'The International Regime for Treating Civil Wars, 1988 2017', *Daedalus*, vol 147, no 1.

Gowan, R. and Andersen L.R., 2019. 'What next for UN peace operations? Global crisis management in a post-Covid19 world', *DIIS Policy Brief*, Danish Institute for International Studies, Copenhagen https://www.diis.dk/en/research/what-next-un-peace-operations

Guéhenno, J-M, 2018. 'The United Nations & Civil Wars', *Daedalus*, vol 147, no. 1.

Hunt, C.T. and Zimmerman, S., 2019. 'Twenty Years of the Protection of Civilians in UN Peace Operations: Progress, Problems and Prospects', *Journal of International Peacekeeping*, vol 23, no 1-2.

Hunt, C.T., 2020. *Waiting for Peace: A Review of UNMISS' Political Strategy in South Sudan*. United Nations University, New York.

Howard, L.M. and Stark, A., 2018. 'How Civil Wars End: The International System, Norms, and the Role of External Actors', *International Security*, vol 42, no. 3.

Johnstone, I., 2011. 'Managing Consent in Contemporary Peacekeeping Operations', *International Peacekeeping*, vol 18, no 2.

Jones, S., 2014. 'A Persistent Threat: The Evolution of al Qa'ida and Other Salafi Jihadists', RAND Corporation.

Luttwak, E.N., 1999. 'Give war a chance', *Foreign Affairs*, vol 78, no 4.

Maganza, B., 2018. 'Escalation of Violence in Bangui: Has MINUSCA Become Party to a Conflict in CAR, and What Would That Mean?' *Opinio Juris*, http://opiniojuris.org/2018/04/16/escalation-of-violence-in-bangui-has-minusca-become-party-to-a-conflict-in-car-and-what-would-that-mean/

Stearns, J. and Vogel, C., 2017. 'The landscape of armed groups in Eastern Congo: fragmented, politicized networks.' *Kivu Security Tracker (KST)*. Centre for International Cooperation, NYU, New York, http://congoresearchgroup.org/wp-content/uploads/2015/11/The-Landscape-of-Armed-Groups-in-Eastern-Congo1.pdf

Stearns, J., 2016. 'Is MONUSCO Doing a Good Job of Protecting Civilians,' Global Peace Operations Review. https://peaceoperationsreview.org/commentary/is-monusco-doing-a-good-job-at-protecting-civilians/

Tinti, P., 2022. Whose Crime Is It Anyway? Organized Crime and International Stabilization Efforts in Mali, Global Initiative Against Transnational Organized Crime, March.

Twagiramungu, N., Duursma, A., Berhe, M.G., and de Waal, A., 2019. 'Re-Describing Transnational Conflict in Africa', *The Journal of Modern African Studies*, vol 57, no. 3.

UN, undated. 'Action for Peacekeeping', Homepage, https://peacekeeping.un.org/en/action-for-peacekeeping-a4p

UN, 1993. An Agenda for Peace: Preventive Diplomacy, Peacemaking and Peace-keeping *Report of the Secretary-general*. United Nations, New York, https://digitallibrary.un.org/record/242288?ln=en

UN, 2000. *Report of the Panel on United Nations Peace Operations*. United Nations, New York, https://www.un.org/en/ga/search/view_doc.asp?symbol=A/55/305

UN, 2015. Uniting our Strengths for Peace: Politics, Partnerships and People. *Report of the High-Level Independent Panel on Peace Operations*. United

Nations, New York, https://peacekeeping.un.org/en/report-of-independent-high-level-panel-peace-operations

UN, 2018a. Declaration of Shared Commitments on United Nations Peacekeeping Operations, A/72/110 – S/2018/815. NB: A4P has since been followed up by A4P+. See: https://peacekeeping.un.org/en/a4p-action-peacekeeping-priorities-2021–2023

UN and World Bank, 2018b. *Pathways for Peace: Inclusive Approaches to Preventing Violent Conflict.* World Bank, Washington, DC, https://www.worldbank.org/en/topic/fragilityconflictviolence/publication/pathways-for-peace-inclusive-approaches-to-preventing-violent-conflict

UN, 2019. 'Action for Peacekeeping: Highlights of Key Achievements', New York: Information Management Unit, DPPA-DPO, https://peacekeeping.un.org/sites/default/files/191101_a4p_achievements_one_pager.pdf

UN, 2022. *Our Common Agenda—Report of the Secretary-General.* United Nations, New York, https://www.un.org/en/content/common-agenda-report/assets/pdf/Common_Agenda_Report_English.pdf

UNU, 2017. 'To Build Consent in Peace Operations, Turn Mandates Upside Down', UNU-CPR, New York, https://cpr.unu.edu/publications/articles/to-build-consent-in-peace-operations-turn-mandates-upside-down.html

Walter, B.F., 2017. 'The New New Civil Wars', *Annual Review of Political Science*, vol 20, no 1.

Walter, B.F., Howard, L.M., and Page Fortna V., 2020. 'The Extraordinary Relationship between Peacekeeping and Peace.' *British Journal of Political Science*, vol 51, no. 4.

World Bank, 2011. *Conflict, Security and Development: 2011 World Development Report.* World Bank, Washington, DC.

15 Conclusion

Richard Iron and Damien Kingsbury

The title of this book and its central theme of how wars end raises a series of questions, the first being why ending war is important. This is not merely an intellectual pursuit, because such understanding might allow us to end wars more quickly and with less human and material loss. And, if greater understanding about how wars end can be achieved, it might help avoid them in the future.

Assuming that wars will start, however, how and why they end appear to reflect in significant part what they were intended to achieve and how they are conducted. One of the certainties of war is its uncertainty; the 'fog of war' means that, once engaged, its precise movements and moments are only imperfectly understood; tactical and strategic knowledge is never absolute and rarely near absolute. The nature of war is such that it can turn on unforeseen events and, of course, war is rarely conducted with perfect strategic understanding at the outset. Mistakes are made, not least in understanding the enemy. Unforeseen circumstances arise. Plans do not survive first contact with the enemy. Allies and enemies change disposition. Third parties intervene to further their own interests. Domestic politics intrude, frequently in response to the changing fortunes and casualties of war—sometimes to encourage, other times to discourage, compromise to help a war end. All these directly contribute to war's conduct and all help shape how wars end.

Countries rarely go to war simply because of the whims of their political leaders. It is common to ascribe to particular leaders who initiate war a personal evil, oversized ego or insanity, and elements of each of these may indeed be present in some. But even where leaders have initiated war, there has usually been an underlying strategic or economic motive which has either compelled the leader to act aggressively or been used to justify an aggressive act, perhaps acting in tandem with other, more personal, qualities.

In this respect, human agency is the capacity to decide, for war or for peace, but it occurs within a context. This context may reflect the fears and insecurities of a leader, a leadership team or which may, as a consequence of history, be deeply instilled within a people. It may also reflect an assertion of regional hegemony or sphere of influence, and the sense of direct threat if

DOI: 10.4324/9781003317487-20

that hegemony is challenged or lost. So, too, loss of access to or greater need of material goods, be they food, energy, water or raw materials, can be a driver of war.

Untangling whether human agency or structure has primacy in decision making, or precisely how they interact, is fraught, both because of the complexity of the entanglement of motives but also because of what is often competing world views, which ascribe importance differently. But at least attempting to unpack war's drivers, understanding that war is a consequence of a complex of factors, might help us better—if not completely—understand why they occur and, potentially, how they might be ended, or even avoided.

Some wars have, in recent times, been begun by larger powers attempting to assert their will on smaller states, such as Russia in Afghanistan and Ukraine, or the US (and allies) in Vietnam and Iraq. In each case, the war did not proceed as initially planned. It is not clear that these wars would have even commenced if the protagonists had been granted the benefit of hindsight. Had there been a logical reassessment of each, at key points, it is not clear that, once started, they would have been continued; none achieved the objectives the initiating powers initially set for themselves.

As with other great powers, China is also extending its strategic reach. The question is whether it will follow a territorially expansionist strategic path and, if so, face similar challenges. To date, and setting aside the problematic incorporation of Tibet, China's territorial ambitions have been limited to the North and South China Seas. It has not—yet—decided to bring Taiwan back into the larger Chinese fold. Alternatively, China's strategic assertions could collide with those of the US and its partners, now notably expanded to include India—China's main regional rival—Japan and Australia.

Perhaps it is the times that make the leader but, if so, these times do not provide a foundation for Vladimir Putin hoping to emulate Peter the Great, Xi Jinping the great Khan, or for US presidents to be so well situated in world affairs as was Franklin Roosevelt. The aspirations of leaders, for personal or reflected glory, for securing a place in history or simply for being the ablest and most far-sighted servant of their peoples, may take their countries to war. But, in each case, the aspirational goals which they seek are rarely met and almost never met without vast and usually unnecessary suffering.

This then brings us back to why wars are started and, by extension, how they are conducted and how they end. This first part of this book has examined why wars start, and the types of wars that unfold once they are started. The second part of this book looked at a number of cases—not exhaustive nor intended to be—but illustrations of different wars and types of wars that, short of a global total war of survival, might mark the next few decades.

The third section of this book looked at some alternative endings to wars, in particular the roles of third parties and negotiations. In most cases, this returns politics to primacy of place in resolving what are, usually, political failures which allowed or encouraged wars to start. The fourth section briefly recapped some ways forward. In all of this, Clausewitz's famous dictum might be turned on its head: war is often the failure of policy by other means.

Is War Ever Justifiable?

In 2022, Bangladesh and Australia celebrated 50 years of diplomatic relations. The Bangladeshi High Commissioner to Australia made the point that the 1971 Indo-Pakistan War was absolutely necessary: his country would not exist without it and most Bangladeshis do not wish they were still in East Pakistan. The same can be said of Ugandans when discussing the 1979 Tanzanian invasion that toppled Idi Amin; of the Irish when talking of their liberation from Britain; or of many other wars that have achieved goals of liberation, secured national identity and removed a foreign force.

In 1939, the UK and France declared war on Germany, not the other way round: so, although (especially in light of what happened in 1940) it is commonplace to regard the UK as fighting largely a defensive war, in 1939 it was the aggressor, albeit to counter a greater wrong. So, even if no one would go so far as to say that war was 'noble', it can be fought for ethical reasons and perhaps have some outcomes which reflect higher moral principles.

It takes two to make war, however. Typically, one party is an attacker; but the defender also makes a conscious decision: to resist rather than to capitulate. So, for there to be war, there needs to both a Johnson and a Ho Chi Minh, a Bush and a Taliban, a Putin and a Zelenski. In this respect, one may propose there is a duality of aggression and confronting aggression; one unjustifiable and the other justifiable. Yet wars are rarely so simple or morally clear; in many cases, wars are a consequence of a series of reciprocal acts that escalate to the point of armed conflict. So, referring to war as 'justifiable' or not is inaccurate: justifiability is defined by the positions, actions and perspectives of the combatants, not by the war itself.

There is a difference between what might be perceived as an aggressive war, and war conducted in response to aggression. But the level of justification is, of course, dependent on the perspective of the observer. The 2022 invasion of Ukraine is viewed very differently in, for example, Moscow, Kyiv, London, Delhi and Beijing. It may be easy to apply absolute moral judgements, but it is rarely helpful to aiding understanding, even in the Ukraine example.

Nevertheless, perceptions of the justifiability of a war impact how it ends. The situation in Ukraine has highlighted the conflict between encouraging compromise to end the war and the perception of allowing an aggressor to benefit from that aggression. If the issue was confined purely to the war at hand, and if there could be guarantees that there would be no resumption of war, then compromise might be reasonable.

But the world is not often like that. The 2014 Minsk Accords aimed to resolve the Donbas War but Russia launched its assault in 2022. In Sierra Leone, the 1999 Lomé Peace Agreement was similarly ineffective. US capitulation in Afghanistan in 2021 and perceptions of its strategic weakness appear to have provoked or inflamed a number of conflicts around the world including, potentially, that in Ukraine.

Thus, there needs to be a balance between three factors: (a) the awfulness of war and the need to end it as soon as possible with the fewest casualties and least damage; (b) relatively abstract concepts of human rights, justice and impunity; (c) and the practical dangers of rewarding aggression which may encourage not just that aggressor but potential aggressors elsewhere.

This therefore poses the dilemma for peacemakers: should there be pressure on the perceived victim of aggression to accept compromise and appear to reward the aggressor, or help the victim of aggression to combat it and thus, potentially, prolong the war? And what of humanitarian intervention, which argues that a war can—even should, with caveats—be conducted in order to correct a greater wrong. This is alien to many who promote and work towards peace. But, as with the causes of war—how it escalates, why, who is at fault, who is aggrieved—it is entirely dependent on the viewpoint of the observer. Armed intervention, though, may be the most effective way of dealing with aggression and discouraging further would-be aggressors.

Of course, there are other circumstances which are far less clear-cut, including many separatist wars, where identifying the origins and initiators of naked aggression is not easy—the Northern Ireland conflict comes to mind—and is more open to perception and interpretation. And, returning to the examples of Bangladesh, Uganda and Ireland, aggression doesn't necessarily have to be tanks and bombs, but could constitute repressive political, economic, ethnic or sectarian discrimination, domination or repression.

The other principal consequence is the more one party to a war feels it is the victim of egregious aggression, the less it is likely to be prepared to offer compromise to end a war. The anger felt by Ukrainians in 2022 appears to have kept them fighting longer and, the longer the war has gone on, to be less prone to negotiation. Similarly, if the aggression is perceived to be unjustifiable by the wider international community, external pressure on the perceived victim to compromise is likely to be markedly reduced: some Europeans originally counselling Ukrainian compromise, yielding land for peace, are more reluctant to be seen taking this position in the face of increasing levels of Russian military violence.

Towards An Approximate Theory of How Wars End

Because no two wars are alike or end in the same way, there can only be an approximate theory of how wars end. But we can discern, from precedent, that wars generally do end with some mix involving a level of coercion and a measure of compromise, usually on and by both sides. The level of coercion might be so compelling that one side can force a surrender on the other; or two sides might be so evenly balanced that neither can realistically coerce the other, described as a mutually hurting stalemate, and both need to compromise broadly equally to bring a war to a close.

There appear to be three main drivers which influence these two variables—coercion and compromise—and how they contribute to war endings, explored below.

Balance of Military Advantage, Now And In The Future

The scale of military advantage, and hence the level of military coercion one side can wield over another, is a major determinant of how wars end. The higher the advantage over an adversary, the fewer political compromises a combatant may have to make; the lower the military advantage, the more one side may need to compromise its objectives. Equally important is the parties' perceptions of how that military advantage will shift over time. If one side perceives that the military advantage will swing further towards it in the near future, then the incentive is to continue fighting until the position is improved and only then negotiate from a position of greater strength. Such a situation may occur when, for example, one party perceives its adversary as being close to exhausting its military resources (people, weapons, money and the will to fight), but is able to sustain or replace its own losses.

This approximate relationship is shown in Figure 15.1, from the perspective of a single party (and there may be many, as in Myanmar). It is intended to illustrate broadly the importance of military advantage to understanding how a war may end. For example, at the 1995 Dayton peace negotiations, the Serbs had low military advantage and had to make high political compromises; the Croats were the opposite and made fewer concessions.

If both sides are in a position of low military advantage over the other, then the conflict has reached a stalemate and there is space for a mediated settlement with both sides making concessions. If one has higher, but not absolute, military advantage than the other, then we may see a level of coercion or compulsion in that negotiation, with an unevenness in the measure of compromise—where coercion is a psychological manifestation of physical military advantage. The nature of any settlement will, therefore, be at least partly dependent on the comparison of the present and perceived future military advantage between the warring parties.

Internal Politics and Popular Will

All political systems in some way depend on a social contract between rulers and ruled, even if in some cases that contract is very much more compelled than in others. Democracies are more directly and obviously dependent on popular support than autocracies; unpopular foreign wars can result in

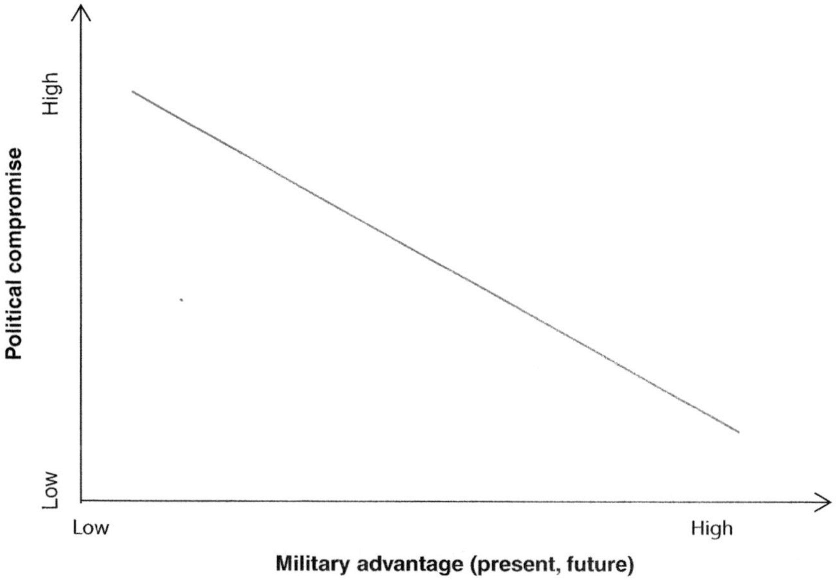

Figure 15.1 Approximate relationship between military advantage and political compromise.

political change at home through the electoral process, sometimes resulting in abrupt changes to war policy including complete withdrawal. But even autocracies, which appear from the outside to be immune to popular pressure, have to be conscious of the threat of a coup or popular uprising. And armies largely reflect the society from which they are drawn: if a population is against a war, it will be difficult to motivate an army (especially a national or conscript army) to fight it sufficiently well to win, even in an autocracy.

So, all parties to a war need a measure of support from their own populations: the consequence being that a population's strength of will in response to a war will impact a party's willingness to continue fighting or consider compromise to stop the fighting. Sometimes the population will be swayed negatively by casualties, such as the USA in Vietnam or Somalia; at other times, the level of casualties or perceived unjustness of the other side's actions might inflame the population and make compromise less likely, with the Soviet Union in World War II, the USA after 9/11 and Ukraine in 2022 as examples. How the war is reported and how the media shape public opinion is, again, frequently dependent on the style of government and its control or otherwise over the media. Even when the media is not government-controlled, it tends to reflect and reinforce popular opinion—for or against war—rather than change it.

War may for one combatant be one of choice, with a particular level of appetite for risk and compromise; but for the other it may be a war of survival, with a very different appetite for risk and compromise. This asymmetry is normal, given it is very rare for the stakes to be the same for both sides. Usually, the side for whom the stakes are higher (for example in the case of national survival) will be prepared to escalate higher and faster, accept higher human and physical costs, take larger risks and be less prone to compromise, than the side for whom the stakes are relatively modest (for example to achieve economic advantage).

Changes to the internal politics of one or more of the combatants may dramatically improve the prospects for a war's end; this may be a result of popular response to the war and how it is being fought. But it may equally reflect a population so inflamed by perceived injustice, or by the need to justify its own sacrifices, that one of the parties to the conflict may be forced to continue fighting or escalate the conflict beyond that which appears, to an impartial observer, to be its rational limits.

But there may also be those who can take advantage from a war's continuation; or who oppose militarily the levels of coercion and compromise that may form the basis for a war's ending. Such spoilers may be small armed factions or splinter groups independent of the main armed forces of either of the parties. An example is the Continuity Irish Republican Army, a splinter group of the Provisional Irish Republican Army, opposed to the Good Friday Agreement and committed to the continuation of violence in Northern Ireland. Alternatively, spoilers may be the military itself, in states where armies are partially or completely independent of state policy, such as the Tatmadaw in Myanmar, during the brief democratic interregnum, or the TNI in Indonesia in the early years of its (partial) democratic transition. In each case, spoilers have to be identified and neutralised within the peace process.

International Pressure and Support

Larger states and powerful multilateral organisations are generally more influential than smaller ones in encouraging conflicts to end, in that they have more economic, diplomatic and military capacity to coerce or persuade one or both sides to compromise. In some cases, economic inducements may be important for one of the parties; in others it may be security guarantees or, sometimes, the threat or use of overwhelming external force may be needed to coerce a party to capitulate or compromise. In each case, international pressure can impact the military advantage/political compromise (Figure 15.1), encouraging greater levels of compromise from one or both sides.

International involvement is also frequently essential in enabling the process of ending a war. This can be through providing mechanisms for negotiation (smaller, less involved, states can be more effective than larger ones in this regard, given their greater perceived neutrality), oversight of the disengagement process and guarantees that the terms of any agreement will be observed by the parties. The UN and other multilateral bodies traditionally fulfil these latter roles. These are very different from the role of international coercion and persuasion, although the UN and other multilateral bodies may have to do both through, for example, the UN Security Council's Chapter VII provisions.

External intervention may also make ending wars more difficult. Outside powers may see a national advantage in joining a war, including the option of utilising it as a proxy war against another interested but external party. So, for example, whereas the participants in the Yemen War, in the early stages, may have been open to tribal reconciliation, it now involves larger sectarian and regional power struggles between Saudi Arabia and Iran that are much more intractable to solve. Similarly, foreign-funded mercenaries in Libya obstruct peacemaking efforts there. This is partly because their use reduces the incentive for peace, compared to if one's own people are fighting and dying, and partly because foreign states now have a stake in the outcome of the conflict which may complicate peace efforts.

There may also be international spoilers which see benefits in war's continuation. Examples are the external funding of mercenaries in Libya or the actions of Charles Taylor who encouraged continuation of war in Sierra Leone to further profit in the trade of blood diamonds. Again, spoilers have to be identified and neutralised within the peace process.

Approximate Theory in Practice

How these three factors interact, to alter the degree of both coercion and compromise needed to end a war, varies in each case. To show the range of options and how this approximate theory operates in practice, below are descriptions of this interaction for each of the six case studies examined in Section II in this book.

East Timor

In this conflict the TNI, the armed forces of Indonesia, had overwhelming military advantage over the East Timorese resistance. Yet this was not sufficient to grant Indonesian success since, in the end, Indonesia was forced to grant East Timor independence.

The East Timorese resistance had built a strong international support network and, with the TNI's widespread human rights abuses, the collapse of the Indonesian economy and the resignation of President Suharto in 1998 ushering political change, there was internal Indonesian domestic pressure to resolve the East Timor situation. This resolution came through an agreed UN-supervised ballot on self-determination in August 1999.

Despite the Indonesian government seeking a settlement, the TNI acted as a spoiler, embarking on a campaign of violence both before and after the ballot. This compelled the USA, a long-standing supporter of Indonesian policy in East Timor, to apply significant pressure on the Indonesian government to allow deployment of an international force led by Australia, also a long-time supporter of Indonesian policy on East Timor. This intervention neutralized the TNI as a spoiler and led to the end of Indonesian rule in East Timor.

In the end, the stakes for continuing the war in East Timor were just not high enough for the new Indonesian government to bear the cost and withstand international pressure for compromise. For the East Timorese people, the stakes were sufficiently high for them to withstand intense military coercion and human rights abuses to achieve their goal of independence.

Sierra Leone

The military balance was more prominent in the Sierra Leone War than in East Timor. In each of the four attempts to broker peace, the Revolutionary United Front only even considered negotiation when facing military adversity. The first three attempts failed because the rebels believed that military advantage would swing towards them during the ceasefires occasioned by peace talks. Only when faced with the prospect of greater military defeat, in the absence of a settlement, did they seriously engage with the peace process.

Changes to Sierra Leone's internal politics helped bring peace. The reintroduction of democracy midway through the war granted a high level of legitimacy to the government, both domestically and internationally. From that point, the government secured significant popular support in the country and had a clear mandate to secure peace through either continued war or negotiation. The international community was appalled by the violence but, tied down in the Balkans, western states were unable to do much except pressure the Sierra Leonean government to compromise further to achieve peace. West African states, in particular Nigeria, were much more clear-eyed about the Revolutionary United Front and its backer Charles Taylor, and did much to help President Kabbah's government win military advantage. Furthermore, international efforts to control the trade in illicit blood diamonds undermined the economy of war throughout West Africa and helped neutralize Taylor's influence.

The main potential spoiler of peace negotiations, in addition to Charles Taylor, was another rebel group which had mutated into the West Side Boys and engaged in seemingly senseless and drug-fuelled violence. Fortuitously, the West Side Boys were largely destroyed by British special forces as part of a hostage rescue operation, so this potential spoiler never interfered with the peace process.

The stakes for both sides were very different: for the government, it was about the survival of the state as a working institution and protector of its people from extreme violence. For the rebels it was about avarice, empowerment and being accustomed to a rebel way of life. As a result of this imbalance, once the government had won a decisive military advantage and the rebels had been coerced into negotiation, peace came quickly and continues to the present day.

Libya

Spoilers have played a much larger role in the Libyan conflict. Both the internationally recognized government in Tripoli and its adversary, the self-styled Libyan National Army in the east, have recruited large numbers of mercenary fighters. Initially from sub-Saharan Africa, mercenary groups have also been sent from Russia and Syria, funded variously by the UAE, Turkey and Moscow.

The use of mercenaries enabled both sides to continue fighting for longer, at a higher intensity and with more lethal equipment than would have been possible with Libyans alone. The war also adopted some of the characteristics of an international proxy war, with both Russia and UAE funding pro-Assad Syrians to fight on behalf of the secular, but authoritarian, Libyan National Army; and Turkey funding anti-Assad Syrians to fight on behalf of the more Islamist Tripoli government.

With this level of support, neither side has appeared to be able to achieve decisive military advantage over the other; but, at the same time, the use of externally-funded mercenaries reduced the cost of continuing the war on both sides. The stalemate has not hurt enough to induce either side to negotiate peace. Even if it did, the involvement of external states would complicate any internal negotiation.

For both Libyan parties the stakes appeared broadly equal—the right to rule a unified Libya and impose a particular kind of ideology upon its people—but, for the external states, the stakes were more complicated, with varying levels of ideology and (for Russia at least) attempts to build a client state in a strategic Mediterranean location. All this means that peace in Libya has been very difficult to find.

Sri Lanka

There were no such spoilers in Sri Lanka where, after nearly four decades of on-off war, Sri Lankan government forces eventually destroyed the armed forces of Tamil separatists and snuffed out attempts at Tamil independence or self-determination.

In Sri Lanka, the absence of direct international involvement enabled the defeat of Tamil armed resistance. India, previously supportive of Tamil attempts at independence, lost patience with the Tamil movement during its intervention on the island in 1987-90, especially angered over its assassination of Rajiv Gandhi, and felt a need to counter China's strategic influence over the Sri Lanka government. Further, the Tamils' widespread use of violence against civilian targets, including suicide bombings, cost it international support. Thus, when under sustained military attack, the Tamils could not rely on external sources of weapons, training or money with which to replace losses, whereas government forces could rely on a steady importation of arms from China, tacit support from India including blockading Tamil supplies, and support in the UN Security Council from Russia and China which could be relied upon to veto any attempt at a humanitarian intervention.

In the early 2000s the military situation had been more balanced and, as a result, a ceasefire had been agreed with *de facto* acknowledgement of Tamil autonomy by the Sri Lankan government. But a newly elected, more hard-line government returned the country to war and, with the prospect of improving military advantage, set about achieving an unconditional military victory.

The stakes for the Tamils were independence or self-determination after what was perceived as unfair Sinhalese domination since independence; for the government, it was the unity of the state. This disparity did allow for compromise, as witnessed in the 2002 ceasefire and recognition of a Tamil area of control; but this compromise was undone by changing internal Sri Lankan politics and the perception of greater future military advantage.

The Sri Lankan war is an example where military advantage and the international isolation of one side enabled the other to impose a complete surrender—or total coercion—on a party for whom the stakes were higher, seemingly had high levels of support from its own population, and was prepared to bear greater human and physical cost. But the Tamils could not survive total military defeat.

Aceh

Although, in some respects, the Aceh conflict has some similarities to the one in Sri Lanka, the outcome was very different with the Acehnese and Indonesia both compromising to find a peaceful solution to the war. As in East Timor, the military advantage favoured the TNI, the armed forces of Indonesia. But over 30 years of military operations it had been unable to defeat the Acehnese separatist rebels who were able to recruit freely from their own population, especially during periods of intense TNI operations. Like many other insurgent groups, they just had to survive to be effective. As a result, the military situation was more a stalemate than raw numbers would indicate.

Whereas, in Sri Lanka, the election of a more hard-line government presaged a return to war, in Indonesia political change was reformist, enabling the possibility of local democracy and, thus, genuine political autonomy for Aceh. But the real circuit-breaker was the response to the 2004 Indian Ocean tsunami: large numbers of international aid agencies and media arrived in Aceh, propelling the war onto the world stage and giving new impetus to both sides to negotiate peace. The TNI, identified as a spoiler to negotiations, could not manage the response to the natural disaster alone and the Indonesian government had to accept the presence of foreigners to run the humanitarian operation.

Both sides compromised: the Indonesians granted self-government; the Acehnese accepted something short of full independence. The involvement of the international community was critical, not least in pre-empting any potential spoiling by the TNI. Once again, the stakes were higher for the separatists than for the larger power but, in this case, it meant that the Acehnese separatist forces were able to bear relatively higher costs to keep their objectives alive, enforcing a military stalemate until the right moment for a negotiated settlement.

Afghanistan

Looking purely at the Afghan war since 9/11, there were three main actors: the Taliban, the Afghan Government and the international intervention force—primarily the USA—supporting the Afghan Government and its military. For much of the war, there was little significant military advantage to either side, except during the US surge of 2009-14. Because the surge's end-date was set when it was announced, the Taliban could expect the military advantage to swing back to it in the future, regardless of any US tactical success in the meantime. In the years following 2014, when the US provided support to Afghan forces who did most of the fighting, the military advantage did indeed swing towards the Taliban, albeit incrementally and not decisively until 2021. At this point, peremptory American withdrawal swung the military advantage suddenly towards the Taliban such that the government collapsed in a matter of weeks.

The centralised government of the Republic of Afghanistan was never popular, in particular in rural areas where loyalties were more traditionally based. The Taliban, on the other hand, could exploit both tribal and religious traditions to build a level of popular support that enabled it to withstand military reverses. Pakistan's support for the Taliban was critical in providing safety to Taliban fighters and their families, largely secure from Afghan government or US attack. In the USA, post-2014, the war was not a major issue either way for most Americans. American casualties were low and commitment was contained. Yet, both Trump and Biden administrations were determined to divest themselves of a war that was seen as unnecessary and irrelevant compared to looming competition with China.

The stakes were high for the Afghans—both for the government and the Taliban. But, for the Taliban, cultural and religious factors provided a higher level of motivation. Pakistan sought a compliant, if not client, neighbour that was antithetical to India. The USA perceived its own stakes to be so low as to accept defeat and withdraw when there was little imperative to do so. Yet, the consequences of American withdrawal and what has been interpreted as a display of weakness appear to have had global consequences, not least in Ukraine. It may be that, for the USA, the stakes were in fact the highest of all, but the Administration was blind to them.

Conclusion

There is no single reason why countries, and peoples, continue to go to war, but there are some common reasons, as this book has explored. Yet the unpredictability of war and its innate complexity continue to be disregarded as though an initial aspiration, motive or momentary need provides sufficient clarity of outcome. Yet wars do end and how they end may provide lessons about ending them more quickly and surely and, perhaps, how to avoid them in the first instance.

War is complex and rarely proceeds as first anticipated. The reasons for war, too, are usually deeply layered and overlapping and cannot easily be disentangled. This is all complicated by the will of political leaders and the temper of people in the face of growing tensions. Setting aside absolute victory or loss, and noting that even these outcomes still require some finessing, unpacking these complexities, resolving that which can be resolvedand setting aside and finding neutralizing mechanisms for that which cannot, is how wars can be ended.

Beyond this, paying attention to rising tensions and identifying and resolving issues before they become flashpoints should be the ordinary work of good diplomacy. But diplomacy cannot address every contingency. Recognizing the key trigger points that mark degrees of escalation in such a process and learning how to neutralize them at each stage, and understanding that the costs almost always exceed their benefits is in the first instance, perhaps, the start of learning how to avoid wars—to have policy, not policy by other means.

Much has been written elsewhere of the utility or otherwise of the principles of war. On the evidence of this book, if we were to contribute to this discussion, we would offer a further principle: beware those who promise certainty or rapidity of outcome in war. Occasionally, things do proceed to plan and one's wished-for speedy and total victory comes to fruition. But more frequently it does not. So, when considering the application of military force, it would be wise to think through in advance the options for how it might end: not just if all goes well, but also if it doesn't. Consider the drivers for ending wars: if the military advantage is not quite what was hoped; if the will of your adversary's population is higher than that of your own and if your own government might be imperilled if the war goes badly; how other states will respond as casualties and damage mount and which might be potential spoilers that could upset a tidy peace? All impact how a war ends and it might not end well for whoever initiated it. So think carefully before starting or joining a war: how might it end?

Index

Abuja Agreement 92
Aceh Monitoring Mission (AMM) 119
Act of Free Choice 82
Afghan Air Force 130
Afghanistan 7, 12–15, 17, 23–25, 30, 32, 36, 38–39, 123–124, 127–130, 132–133, 135–137, 139–151, 164–166, 196–197, 222, 229, 237, 239, 247
Afghan National Army (ANA) 126, 148
Afghan National Defence and Security Forces (ANDSF) 148
Afghan National Police (ANP) 126
AFP; *see* Australian Federal Police
AFRC; *see* Armed Forces Revolutionary Command
AIDAB; *see* Australian International Development Assistance Bureau
Allied Democratic Forces 225
American Civil War 33, 144
AMM; *see* Aceh Monitoring Mission
ANA; *see* Afghan National Army
ANDSF; *see* Afghan National Defence and Security Forces
Anglo-Chinese Wars 38
ANP; *see* Afghan National Police
Anti-Machiavel 17
APEC; *see* Asia Pacific Economic Cooperation
Arab-Israeli Six Day War 2
Arabistan 33
Arab revolt 20
Armed Forces Revolutionary Command (AFRC) 89, 98
ARMM; *see* Autonomous Region of Muslim Mindanao
ASEAN Defence Ministers Meeting-Plus (ADMM-Plus) 166
Asian Financial Crisis 80, 81, 205
Asia Pacific Economic Cooperation (APEC) 85
Association of Indonesian Muslim Intellectuals (ICMI) 80
Australia 59, 70–71, 75–76, 79–80, 83, 85, 127, 157–169, 238, 244
Australian Federal Police (AFP) 160
The Australian Institute of International Affairs Victoria 6
Australian International Development Assistance Bureau (AIDAB) 160
Australian Universal Scheme for National and Community Service (AUSNACS) 168
Australia's Department of Foreign Affairs 71
Austro-Hungarian Empire 179
automation 30, 37
Autonomous Region of Muslim Mindanao (ARMM) 195
Awaluddin, Hamid 113
Ayran race 33
Aztec 38

Bangladesh War of Independence 38
Best Alternative to Negotiated Agreement (BATNA) 191
Biafran War 33
Biden, Joe 25
Biden Administration 176
Blair, Tony 24
Blaxland, John 155
Bolshevik Revolution 144
Bourbon authority 17
Brahimi, Alia 57
British Commonwealth Far Eastern Strategic Reserve 159
Bronze Age 32

Cameron, David 24, 25
CAR; *see* Central African Republic
Carnation Revolution 70
Casablanca Conference (1943) 5
CDF; *see* Civil Defence Force
Cease Fire Agreement 116
Central African Republic (CAR) 60, 221
Central Powers 20
Chamberlain, Neville 144
Charter of the United Nations 23
China 2, 20, 121, 155, 174, 178–184, 237, 245–247
Churchill, Winston 15, 144, 147
Civil Defence Force (CDF) 89, 91
civilizing mission 2
Clark, William P 84
Clausewitz, C. von 11, 14–19, 22–23, 26, 29, 39, 46, 172, 204, 238
Clinton, Bill 162
Clinton administration 84
Cohen, Eliot 46
Coker, Chris 30
Cold War: anti-communist atmosphere of 75; 'end of history' 4; fall of the Soviet Union 12; in favor of liberation 2; geopolitical context of 71; strategic contest of the twentieth century 173; US nuclear strategy 176
Communism 22
Comte de Guibert 18
Continuity Irish Republican Army 242
Conventional war 33
counterinsurgency (COIN) 123
Counter Terrorism Service 53
Crisis Management Initiative 6
culture 107, 109, 208; language and 19; popular 49; strategic 30

Dalmatia 20
Darul Islam 109
DDR; *see* Disarmament, Demobilisation and Reintegration
Democratic Republic of the Congo (DRC) 221
Department of Foreign Affairs and Trade (DFAT) 160
Disarmament, Demobilisation and Reintegration (DDR) 99
Doha Agreement 128, 142–143; Asian Financial Crisis 81; human rights violations 77; 'Joint Declaration' 142; settlement with the Taliban 128; US Secretary of State Pompeo 142
Donbas War 239
DRC; *see* Democratic Republic of the Congo

East Timor: Carnation Revolution in Portugal 70; National Press Club 81; Truth Commission 83; United Nations General Assembly resolutions 74; UN Security Council 73; US equipment 72
East Timorese Students and Youth Association (IMPETTU) 76
electro-magnetic spectrum 43
electronic communications 43
Euphrates River 46, 48
European Age of Enlightenment 3
European Union 22, 83, 174

Falkland Islands 12
Fall of Constantinople 3
Fernandes, Clinton 57
Feynman, Richard 146
Five Power Defence Arrangements (FPDA) 163
FPDA; *see* Five Power Defence Arrangements
France 4, 12, 18–20, 29, 35, 61–62, 74, 147, 172–173, 198, 238
Franco-German War 35
First World War 15, 20, 29, 173, 177, 179; *see also* World War I
Frederick the Great of Prussia 17
Free Aceh Movement 107
Free Papua Organization 212
Freetown Peninsula 94
French-Indian Wars 36
French Revolution 3, 16, 19, 21
FRETILIN 70–74, 76

Gandhi, Rajiv 245
Geneva Conventions 3
Germany 2, 4–5, 11–12, 15, 19–20, 23, 33, 147, 157, 173–174, 177, 198, 238
Ghali, Brahim 206
Ghani, Ashraf 5, 133, 197
Global War on Terror 164
Good Friday Agreement 242
Government of Indonesia-Free Aceh Movement (GoI-GAM) 207
Government of Morocco 205
Government of the Republic of the Philippines-Moro Islamic Liberation Front (GRP-MILF) 209

The Grand Alliance 22
Great Britain 32
Great Power club 172
Grenada 6
Guerrillas 38
Gulf War 12, 159

Hati Suci (Sacred Heart) 77
High Frequency Trading markets 40
high intensity 34
High-Level Independent Panel on Peace Operations (HIPPO) 228
HIPPO; *see* High-Level Independent Panel on Peace Operations
Hobbes, Thomas 51, 143
Holy Roman Empire 36
Holy War 20
Honourable East India Company 35
Howard, John 164
The Hundred Years War 35
Hurley, David 160
Hussein, Saddam 14, 33

ICEDS; *see* Institute for Climate, Emergency and Disaster Solutions
ICG; *see* International Crisis Group
Indian Peace Keeping Force 119
Indo-Pakistan conflict 5
Institute for Climate, Emergency and Disaster Solutions (ICEDS) 168
International Coalition 46
International Crisis Group (ICG) 206
International Force in East Timor (INTERFET) 85, 162
International Monetary Fund 231
International Relations 75
International Security Assistance Force (ISAF) 123, 197
Iran-Iraq War 6
Iraq 6, 9, 12–15, 17, 24, 33, 36, 46–54, 140, 159, 165, 222, 237
Iraqi National Security 52
Iraqi Security Forces 46, 47, 49, 51, 53
Iraqi State 47–48, 54
Iron, Richard 57
ISAF; *see* International Security Assistance Force
Islamic Republic of Afghanistan 165
Islamic State 6

Jakarta 159
Japan 2, 4–5, 20, 30, 75, 157, 173–174, 179, 237
Johnson, Boris 25
Johnson, Dominic 37
Johnson, Robert 9, 57
Joint Defence Facility 158

Kaldor, Mary 30
Kantian peace theory 31
Karuna faction 118
Karzai, Hamid 14, 140
Khan, Ismail 134
Kilinochchi 111
Kilcullen, David 58
Kingsbury, Damien 58
Kohen, Arnold 74, 75, 80
Korean Armistice Agreement 5
Korean War 5, 158
Kosovo 6, 162

Land Component Commanding General 48
League of Nations 21, 23, 173, 194
liberal capitalism 22
Liberation 29
Liberation Tigers of Tamil Eelam (LTTE) 107, 212; *see also* Tamil Tigers
Libya 6, 12, 14, 17, 24, 57, 59–66, 111, 195, 223, 243, 245
Libyan civil war 59
Libyan National Army (LNA) 59, 65, 245
LNA; *see* Libyan National Army
Lomé Peace Agreement 95, 239
Louis XIV of France 17
Low intensity 34
LTTE; *see* Liberation Tigers of Tamil Eelam
Luttwak, Edward 222

Majapahit Empire 108
Maley, William 58
Massachusetts Institute of Technology 75
May, Theresa 25
McArthur, Douglass 54
McCormick, Gordon 132
McNaughtan, Andrew 81
MHS; *see* Mutually Hurting Stalemate
Middleton, Karen, *An Unwinnable War* 165
MILF; *see* Moro Islamic Liberation Front
Minister for Research and Technology 81

Morgenthau, Hans, *Politics Among Nations* 158
Moro Islamic Liberation Front (MILF) 6, 195
Moro National Liberation Front (MNLF) 195
Mutually Hurting Stalemate (MHS) 189–190

Nagorno-Karabakh 30
Napoleonic Wars 15, 19, 36
National Capitals 48
National Commission of Inquiry 78
National Federal Republic of West Papua 212
national institute for net assessment (NINA) 168
National Parliament of West Papua 212
National Press Club 81
National Resistance of East Timorese Students (RENETIL) 77
NATO 6, 22, 61, 63, 148, 174–177, 198
Nazism 15
New Zealand Defence Force 161
Nobel Peace Prize 80
Noble, Roger 9
non-state armed groups (NSAGs) 192
NSAGs; *see* non-state armed groups

Obama, Barack 15, 25, 123
Opala, Joseph 104
Operation Barbarossa 21
Operation Opus 59
Operation Sabotage Elections 96
Opium Wars 33
Organization of Islamic Conference 73
Ottoman Empire 15, 20, 38, 173
Our Common Agenda 232

Pacific Island Forum 166
Pan-German League 20
Pattison, James 64
Pearl Harbor 24, 33, 147
Peninsular War 19
PFLOAG; *see* Popular Front for the Liberation of the Occupied Arabian Gulf
Pinker, Steven 30
political realism 39
Popular Front for the Liberation of the Occupied Arabian Gulf (PFLOAG) 5
Popular Mobilisation Forces 53
power: accumulation of 9, 28, 43; civil 19; ex-colonial 100, 104; latency of 43; military 13, 34, 91, 157, 166, 169, 174; national 158, 161, 165, 167–169; regional 6; relative 43, 173; resources and 40; wealth and 173, 183; world 5
Protection of Civilians (PoC) 225
Provisional Irish Republican Army 242
Putin, Vladimir 24, 174, 198, 237

radio communication 33
Red Force 49
Republic of Sierra Leone Armed Forces (RSLAF) 100
Responsibility To Protect (R2P) 23–24, 206
Revolutionary and Napoleonic Wars 4
Revolutionary Armed Forces of Colombia (FARC) 193
Revolutionary United Front (RUF) 89, 244
Revolution in Military Affairs 159
Roman Empire 36
Romanov regime 21
Roosevelt, Franklin Delano 21, 237
Royal Australian Air Force (RAAF) 157
Royal Australian Navy (RAN) 157
RSLAF; *see* Republic of Sierra Leone Armed Forces
RUF; *see* Revolutionary United Front
Rumsfeld, Donald 164
Russian Civil War 144

Saddam Hussein 12
SADF; *see* South African Defense Force
Santa Cruz 77
Schroeder, Paul 172
Scott, James C. 149
Second World War 15, 17, 21–23, 32, 35, 144, 157, 173–174, 177; *see also* World War II
Security Council 6
security force assistance (SFA) 123
self-determination 3
Senkaku/Diaoyu Islands 2
SFA; *see* security force assistance
Shaka Zulu's war 34
Sierra Leone 6, 57, 64, 89–93, 95–96, 98–105, 239, 243–244

SLMM; *see* Sri Lanka Monitoring Mission
Smitham, Thomas D. 60
social behavior 1
SOF; *see* Special Operations Forces
South African Defense Force (SADF) 194
Soviet Union/USSR 4, 5, 12, 20, 21, 33, 38, 182, 241
Special Operations Forces (SOF) 125
Sri Lanka 5–6, 58, 107–109, 116–121, 212, 245–246
Sri Lanka Armed Forces 112
Sri Lanka Freedom Party 117
Sri Lanka Monitoring Mission (SLMM) 119
Stalin, Joseph 3
Stavanger Declaration 111
stakeholders 46–47, 49, 53
Strachan, Sir Hew 9
Sultanate of Zanzibar 32

Tamil Tigers 5; *see also* Liberation Tigers of Tamil Eelam
Taylor, Charles 89, 94, 101, 243–244
Tehran 33
Tentara Nasional Indonesia (TNI) 111
Tigris River 46
Timorese Council for National Reconciliation (CNRT) 213
Timor-Leste 6, 86n2, 169n1, 212–213
Tim Pembina Desa (TPD) 73
TNI; *see* Tentara Nasional Indonesia
Tora Bora Mountains 164
TPD; *see* Tim Pembina Desa
Tragedy of War, The: European Age of Enlightenment 3; Fall of Constantinople 3; French Revolution 3; Geneva Conventions 3; Mongol Siege of Baghdad 3
Train, Advise and Assist (TAA) 124
Tyrannies 34

Ukraine 2, 6, 22, 25, 60, 62–63, 65, 150, 167, 174–176, 180, 198, 200, 232, 237–239, 241, 248
UNAMET; *see* United Nations sponsored Assistance Mission to East Timor
UN Charter 24, 25
UN General Assembly 63, 74
United Kingdom 17, 25, 59, 89, 99, 108, 147, 159, 163
United Liberation Movement for West Papua 212
United National Party 117
United Nations sponsored Assistance Mission to East Timor (UNAMET) 162
United States (US) 12–14, 17, 20–23, 25, 34, 59, 70–72, 74–76, 83–84, 99, 108, 124, 133, 136–137, 139–143, 145–150, 157–159, 162–163, 165, 174, 181–183, 210; naval base 2; policy 75, 84, 180
Unmanned Air Vehicles 30
UN Security Council 73, 100, 121, 229–230
UN Special Envoy and the Personal Representative of the Secretary-General (SE/ PRSG) 196
US Department of Defense 61
US Secretary of Defense 85
US-Taliban agreement 143
UN Mission for the Referendum in Western Sahara (MINURSO) 205
UN Secretary General 74
UN Security Council 144, 246

Vietnam 22, 159, 237, 241
Vietnam War 24
von Bismarck, Otto 11
von Bülow, Heinrich Dietrich 18
von Clausewitz, Carl 11
von Moltke, Helmuth 11

Wagner Group 60, 65
War of Russian Aggression 198
Wars of German Unification 19
Warsaw Pact 4
Washington, George 37
West Papua National Coalition for Liberation 212
White, Hugh 155
Why War? 1–3; The book 6–7; 'civilizing mission' 2; 'de-Nazification' 2; How Some Wars Have Ended 4–6; 'hurting stalemate' 6; scourge of our Earth 1; Taliban's 2021 victory 6; The Tragedy of War 3–4; 'unconditional surrender' 5; US involvement in World War I 2
Wilson, Woodrow 2, 20, 21

Wilsonianism 21
World Food Program 160
World War I 2, 4, 207; *see also* First World War
World War II 4–5, 75, 86, 155, 193, 241; *see also* Second World War
World War Three 175

Yemen War 243
Yom Kippur War 6

Zartman, William 187